Internet of Things

Technology, Communications and Computing

Series Editors

Giancarlo Fortino, Rende (CS), Italy

Antonio Liotta, Edinburgh Napier University, School of Computing, Edinburgh, UK

The series Internet of Things - Technologies, Communications and Computing publishes new developments and advances in the various areas of the different facets of the Internet of Things. The intent is to cover technology (smart devices, wireless sensors, systems), communications (networks and protocols) and computing (theory, middleware and applications) of the Internet of Things, as embedded in the fields of engineering, computer science, life sciences, as well as the methodologies behind them. The series contains monographs, lecture notes and edited volumes in the Internet of Things research and development area, spanning the areas of wireless sensor networks, autonomic networking, network protocol, agent-based computing, artificial intelligence, self organizing systems, multi-sensor data fusion, smart objects, and hybrid intelligent systems.

Indexing: *Internet of Things* is covered by Scopus and Ei-Compendex **

Lidia Fotia • Fabrizio Messina • Domenico Rosaci •
Giuseppe M. L. Sarné
Editors

Security, Trust and Privacy Models, and Architectures in IoT Environments

Springer

Editors
Lidia Fotia
University of Salerno
Salerno, Italy

Fabrizio Messina
University of Catania
Arcavacata, Italy

Domenico Rosaci
University of Reggio Calabria
Reggio Calabria, Italy

Giuseppe M. L. Sarné
University of Milano-Bicocca
Milano, Italy

ISSN 2199-1073 ISSN 2199-1081 (electronic)
Internet of Things
ISBN 978-3-031-21939-9 ISBN 978-3-031-21940-5 (eBook)
https://doi.org/10.1007/978-3-031-21940-5

This Springer imprint is published by the registered company Springer Nature Switzerland AG
The registered company address is: Gewerbestrasse 11, 6330 Cham, Switzerland

Preface

In recent years, the increasing number of IoT devices, varying from wearable and personal devices to industrial and environmental monitoring sensors, has induced a rapid development of several computing paradigms like fog and edge computing, and approaches often based on distributed architectures and artificial intelligence techniques. All this has stimulated the needs for applications and services in different domains to offer services for smart homes and cities, smart grid, smart community, e-commerce, and many other applications. Nevertheless, while benefits of IoT are widely acclaimed, it is also evident that the necessity of providing security and privacy has emerged as a prominent concern, which considerably inhibits the further developing and deployment of IoT infrastructure and services.

In particular, an IoT infrastructure, potentially consisting of an overwhelming amount of (heterogeneous) devices, can be extremely complex and expensive to protect from malicious attacks, because even a single intrusion success in an IoT device can compromise even the whole infrastructure, making it attractive for cybercriminals. In addition, IoT also provide appealing services to humans, and many researches have shown how such services can be most fruitfully leveraged by fostering cooperation among smart objects. This has pointed to the importance of carefully selecting the best partners to cooperate with, mainly in those scenarios where IoT devices often move between multiple federated environments, where it is difficult to reference the trustworthiness of IoT objects. In such a scenario, the development of a model of trust and reputation, and of successful strategies to recognize the agent's trustworthiness, is becoming a central question in the IoT environment, which is the reason why many approaches have been proposed to equip smart IoT objects with strategies to trust counterparts and spread their trust value to each other in form of suggestions or with some "word of mouth" approaches, thus enabling an IoT system to be adaptive to the time evolution of social relations.

This book is dedicated to the issues of security, trust, and privacy models and architectures in IoT environments, and aims to capture the latest research and contributions from academy, industry, and other stakeholders on new security models, architectures, protocols, and standards for ensuring trustworthiness to IoT systems. We have been interested in, but not limited to, the convergence of IoT, software agents, and edge computing to introduce social features into IoT systems, combining trustworthiness and reputation information collected by agents at the edge with security and privacy mechanisms. We have also addressed experimental, even simulated, campaigns to evaluate strategies to improve the security and privacy of the IoT world, and at the same time the ability to prevent and deter deceptive behaviors.

Within the framework described above, this book brings together in 10 contributions that provide the reader with an overview on the most recent research and trends in the area of security instances, adoption of trust and privacy models and architectures in IoT environments. More in detail, in:

- **"IoT Network Administration by Intelligent Decision Support Based on Combined Neural Networks,"** by Igor Kotenko, Igor Saenko, and Fadey Skorik, deals with the issue of the effective administration of large and complex IoT networks. To this aim, a dedicated intelligent decision support system is proposed for inclusion in the tools of network administrations. The chapter describes the structure of a combined neural network capable of solving the problem of assessing the state of computing network elements, considering three training approaches.
- **"A Novel Privacy-Preserving Framework Based on Blockchain Technology to Secure Industrial IoT Data"** by Aitizaz Ali, Mehwish Kundi, Mehmood Ahmed, Ateeq ur Rehman, and Hashim Ali.
  IoT finds application in the healthcare sector. To avoid security problems in that area, when designing a distributed model, one possible solution is to provide for the adoption of blockchain features in an IoT infrastructure. Blockchains provide tamper-proof and data non-repudiation features. In this proposal, the hyperledger fabric was used to realize the blockchain through chaincode and then we used the cipher for threat and security check. The results were compared with reference models verifying that the proposed solution provides better privacy preservation.
- **"Detecting Collusive Agents by Trust Measures in Social IoT Environments: A Novel Reputation Model"** by Mariantonia Cotronei, Sofia Giuffrè, Attilio Marcianó, Domenico Rosaci, and Giuseppe M. L. Sarnè. This chapter addresses agent-based IoT social contexts, where to mitigate the risks of interacting with unreliable partners exposing agents to risks from malicious behavior, trust, and reputation information are considered. In particular, two main contributions are proposed. The first one consists of a method to preliminarily identify the best candidates as malicious in order to consider them as untrustworthy entities. The second one is a new effective reputation model that can detect collusive malicious agents without introducing side effects with respect to the reputation scores of honest agents, unlike other well-known trust and reputation systems.
- **"An Adaptive Blurring Routing Protocol for Delay Tolerant Networks in IoT Environments"** by Danilo Meli, Filippo Luigi Maria Milotta, Corrado Santoro, Federico Fausto Santoro, and Salvatore Riccobene.

To improve the performance of IoT networks from a security perspective, the authors of this chapter propose an innovative routing protocol that can handle sensitive information, with a delivery performance comparable to the currently most widely used protocols. The simulations carried out show good performance in reaching the correct destination.

- "**Modeling and Detection of Denial-of-Sleep Attacks on Autonomous IoT Devices in Wireless Sensor Networks**" by Vasily Desnitsky.
  Modeling and detection of denial-of-sleep attacks in wireless sensor networks is addressed in this chapter. These efficient and stealthy attacks can reduce or consume the battery life of a device, and thereby deactivate it. Machine learning techniques are proposed to detect denial-of-sleep attacks. Tests conducted have shown to demonstrate the effectiveness and applicability of the proposed technique.
- "**Machine Learning Methodologies for Preventing Malware Obfuscation**" by Vincenzo Carletti, Alessia Saggese, Pasquale Foggia, Antonio Greco, and Mario Vento.
  There is a consistent competition among malware detection systems, but an outstanding question is how robust machine learning-based systems are against obfuscation techniques. In this chapter, seven trained methods for classifying malware are compared, with a focus on recent image-based approaches. The SOREL-20M, one of the largest malware public datasets, was used to perform the comparison, and four obfuscation methods were used. All the methods tested proved to be effective on unmodified test samples, but only a few of them proved to be robust compared to the obfuscation techniques considered.
- "**Formation of Reliable Composite Teams for Collaborative Environmental Surveillance of Ecosystems**" by Giancarlo Fortino, Lidia Fotia, Fabrizio Messina, Domenico Rosaci, Giuseppe M. L. Sarnè, and Claudio Savaglio.
  The authors propose a trust-based architecture for dynamically forming teams in which humans and heterogeneous IoT devices cooperate for environmental surveillance of ecosystems. To form good teams denoted by high levels of mutual trust among its members, this chapter defines a trust measure taking into account both the reputation and measurement accuracy of devices and proposes a framework that, based on such a trust measure, forms the best possible temporary teams at a given time. The results of some simulations confirmed the benefits of the proposed approach.
- "**Accountability of IoT devices**" by Angelo Furfaro, Carmelo Felicetti, Domenico Saccà, and Felice Crupi.
  The use of IoT devices poses critical issues related to security, privacy, and protection of the data on which they operate. In this context, the accountability of IoT devices in the processing tasks they perform in interacting with other systems is relevant. The integration of physical unclonable functions technologies with a DLT-based solution is exploited to address the ability of IoT devices to reliably authenticate and identify the parties involved and the traceability of the operations performed. Described here is the architecture of the proposed infrastructure

that is based on blockchain technologies that underpin decentralized autonomous organizations.

- "**Digital Twin Through Physical Assets Tokenization in Blockchain**" by Giuseppe Ferrara and Corrado Santoro.
  More and more proposals consider IoT and blockchain technologies together. In this chapter, in order to leverage the advantages of tokenized assets in IoT, some relevant aspects of the main approaches of available tokenization approaches are first analyzed, and a new concept architecture is proposed to combine the best sides of each approach into a new hybrid solution. It uses on-chain and off-chain transactions to be suitable for any type of physical asset even in the IoT environment.
- "**Security in Home Automation**" by Julian Calleja and Lalit Garg. The increased and growing prevalence of home automation systems can potentially be the source of security and privacy issues. In this chapter, an overview on the state of the art and future challenges is provided.

We highlight that the chapters published in this volume tackle a wide variety of topics related to security and trust in IoT environments, providing a useful overview of this currently emerging research field. We therefore hope that this book can serve as a reference for a broader audience such as researchers and teachers, application designers and solution architects, but also practitioners, IT managers, and decision-makers.

Salerno, Italy — Lidia Fotia
Arcavacata, Italy — Fabrizio Messina
Reggio Calabria, Italy — Domenico Rosaci
Milano, Italy — Giuseppe M. L. Sarnè

Contents

IoT Network Administration by Intelligent Decision Support Based on Combined Neural Networks

Igor Kotenko, Igor Saenko, and Fadey Skorik

1 Introduction

Currently, the Internet of Things (IoT) networks are widely used in modern network infrastructures. In addition to home networks and "smart home" [31], the spheres of implementation of IoT networks are industry [14], robotics [23], energy [34], mobile communications [2], transport [7], medicine [41], agriculture [10], sports [13], and other areas, where it is possible and necessary to integrate remote electronic devices and traditional computer networks and systems. For this reason, IoT networks are considered as a kind of cyber-physical systems [25]. At the same time, remote network elements ("things") have the following features: (1) they have built-in computer systems, which, in turn, have low resource consumption (low performance and small amounts of memory), (2) they have to process large arrays of heterogeneous data, (3) they are distinguished by a great variety and a large number of developers, (4) they have a sufficiently developed and branched structure, and (5) they can have both a wide purpose and be specific, highly specialized [28, 35]. In addition, modern IoT networks are characterized by the ability to easily and quickly change their structure over time, expand or contract, and include new tools or exclude them from the network. Moreover, this trend of dynamic restructuring and expansion of IoT networks will continue for the foreseeable future in almost all areas of their application [27]. In addition, there are many problematic issues in IoT networks, concerning, in particular, their security [12, 32, 43].

I. Kotenko (✉) · I. Saenko
St. Petersburg Federal Research Center of the Russian Academy of Sciences (SPC RAS), St. Petersburg, Russia
e-mail: ivkote@comsec.spb.ru; ibsaen@comsec.spb.ru

F. Skorik
St. Petersburg communication academy, St. Petersburg, Russia
e-mail: fa_sk@bk.ru

L. Fotia et al. (eds.), *Security, Trust and Privacy Models, and Architectures in IoT Environments*, Internet of Things, https://doi.org/10.1007/978-3-031-21940-5_1

The high dynamics of the behavior of IoT networks with large volumes of processed information and transmitted traffic, as well as other features inherent in IoT networks, cause certain difficulties in solving the problems of administration of such networks [39]. Such tasks, in particular, include monitoring the state of elements of IoT networks, assessment of emerging network incidents according to their degree of criticality, making decisions on restoring the information process or increasing network bandwidth, etc. [36, 38]. Problems such as the organization of polls and the determination of trends in changes in the states of individual devices operating in a heterogeneous environment, the receipt and processing of data coming from them, and the implementation of control services and applications do not currently have unambiguous answers. Moreover, the complexity of solving these problems increases with an increase in the dimension of the IoT network. The larger the size of the computer network, the more complex the administration processes. Accordingly, the labor costs of network administration increase, as well as the need for qualified system administrators. Such needs are often very difficult to meet, especially in small and medium-sized enterprises, where it is not always possible to introduce additional administrative positions in a timely manner (there is a problem of lack of administrative staff), and the load on the network elements tends to constantly increase.

There are a number of ways to tackle the problem of network administrative staff shortages. One of the directions is an attempt to use the services of network outsourcing, which are provided by some specialized providers. However, this is not always possible for financial or information security reasons. Another direction, which is considered in the chapter, is an attempt to reduce the load of network administration by introducing a specialized intelligent decision support system for network administrators into the arsenal of network administration tools. This system receives data on the current state of IoT network elements as input, analyzes them in real time, and develops solutions for managing the IoT network in order to preserve its functioning parameters. The key stage in the operation of such an intelligent system is the timely and complete processing of data on the state of IoT network elements. The faster and more completely such processing takes place, the more correct recommendations and warnings are issued to the IoT network administrator.

As a rule, such decision-making systems rely on some kind of ready-made knowledge base. However, in this case, they cannot go beyond the information available in these databases. This imposes a number of restrictions on such systems. In particular, such systems are not capable of independent development and are not able to quickly and adequately rebuild in the event of an IoT network reconfiguration. Therefore, one of the possible solutions to this problem is seen in the introduction of a specialized analytical unit into the decision support system used by the IoT network administrator, which is based on artificial neural networks (ANN). The presence of an ANN in such a system will greatly simplify the analysis of the incoming data. In addition, this will make it possible to automate the identification of emergency situations, as well as the prerequisites for them without human intervention [9, 21, 33, 42].

The investigation of the possibilities for constructing and evaluating the efficiency of such a system is the main purpose of this chapter. The chapter considers the model of the analytical block of the intelligent decision support system used by network administrators, which is based on the combined ANN, as well as the results of its testing based on various methods of the ANN training. The initial data for the analysis is a normalized set of values of indicators that determine the state of the elements of a computer network at the current time. Based on the neural network analysis of data on the state of network elements, an assessment of the state of the IoT network as a whole is formed [33].

The theoretical contribution of the chapter lies in the development and use of the original model of the analytical block for the intelligent decision support system intended to be used by IoT network administrators, which implements the elements of artificial intelligence and is based on a combined ANN. In addition, the proposed architecture of the analytical unit was tested using known training methods, which resulted in recommendations to apply these methods in the work of the analytical unit.

The results of this research were partially reported in [20]. Compared to [20], this paper was considerably reworked:

- The statement of the research problem has been clarified.
- The number of analyzed related works has been increased.
- The general architecture of the analytical unit and the architecture of each module that is part of it have been developed and specified.
- The theoretical foundations are considered in more detail in terms of describing the work of the ANN module.
- Broader experimental studies have been realized; they, firstly, cover not only ANN training but also testing and, secondly, investigate ANN at higher dimensions.

The further structure of the chapter is as follows. Section 2 presents an analysis of the related work. Section 3 discusses the formulation of the problem of intelligent decision support based on combined neural networks and presents the main indicators characterizing the IoT network state. The general architecture of the analytical block for the intelligent decision support system intended for usage of network administrators, the architecture of its components, and the functioning algorithm of the proposed combined neural network are presented in Sect. 4. Section 5 discusses implementation issues and experimental results for evaluating the proposed system, including a comparative assessment of the error for various training methods. Section 6 contains the main conclusions and directions for further research.

2 Related Work

A fairly large number of works have been devoted to the use of ANN in monitoring systems and intelligent decision support.

In [4], to support decision-making, it is proposed to use an expert system based on ANN and fuzzy logic rules. The core of the expert system is the knowledge base. Knowledge is the rules for interpreting input data. The advantage of this solution is the good stability of the system. At the same time, it should be noted that this system did not have the ability to self-learn. For this reason, its application for solving network administration problems seems to be extremely limited.

The use of pretrained neural networks has been studied extensively. For example, [24] proposed a method for implementing this type of neural networks for modeling the financial performance of a company. Such neural networks have a high learning rate. At the same time, this solution is characterized by a narrow scope of their application, which does not allow them to be fully recommended for network administration.

Fast learning and low need for computing resources are inherent in a special type of neural network, which are Hamming networks. In [16], an original algorithm for classifying real objects based on the use of such networks is proposed. However, a significant drawback that makes it difficult to use Hamming networks for network administration is their low resistance to noise and distortion of input data.

Hybrid neural networks are more resistant to noise. In [26], a self-learning system based on hybrid neural networks and rules of fuzzy logic is considered. However, the disadvantage of this system is the need for significant computational costs for its training.

Among the original decisions on the use of ANN in decision-making systems should be attributed the work [15], in which it is proposed to use bioinspired algorithms for ANN training, namely, the particle swarm algorithm. Despite the originality of this solution and the high stability of learning achieved with its help, it should be recognized that this solution is effective only for small dimensions. With an increase in the dimension of the ANN, the complexity of solving this problem increases sharply.

A significant reduction in the training time and the size of the training sample can be achieved by using a combination of probabilistic and multilayer neural networks, as was investigated in [21]. However, in this case, the possible error depends very much on the noise level in the network.

Many solutions related to the use of ANN are focused on detecting network attacks. So, in [8], a method of using adaptive resonant neural networks is described, which makes it possible to detect and classify certain types of network attacks. At the same time, it should be noted that, despite the promising nature of such neural networks, this method requires laborious training and needs improvement. In [5], an infrastructure method for detecting DDoS attacks based on the Bayesian model of multiple changes is proposed. Its advantage is a fairly high accuracy of calculations. However, this method has a limited area of use. The possibility of using neural networks to determine the possible actions of users in social networks was studied in [29]. This work uses machine learning methods, in particular, a simple neural network. The disadvantage of this solution is the complexity of the analysis and reduction of the source data to a numerical form. Finally, it is worth highlighting the studies in which the self-organizing Kohonen map was analyzed, in particular, the work [40]. The advantage of this solution is the ability to process information with a high level of noise. However, the disadvantage is the difficulty in interpreting the results of the work.

Note that the idea of using the Kohonen map is embedded in the combined ANN proposed in this paper. This idea was further developed in the paper.

In general, based on the analysis of related works, it can be concluded that the known decisions on the use of ANN in decision-making systems, as a rule, have a narrow focus and significant restrictions in the areas of their use.

3 Problem Statement

The place in the network of the intelligent decision support system for the administrator focused on making decisions on network management is shown in Fig. 1.

For the functioning of the proposed intelligent decision support system, two different locations can be provided. The first location position can be associated with a separate dedicated network node. In this case, the system provides, as a

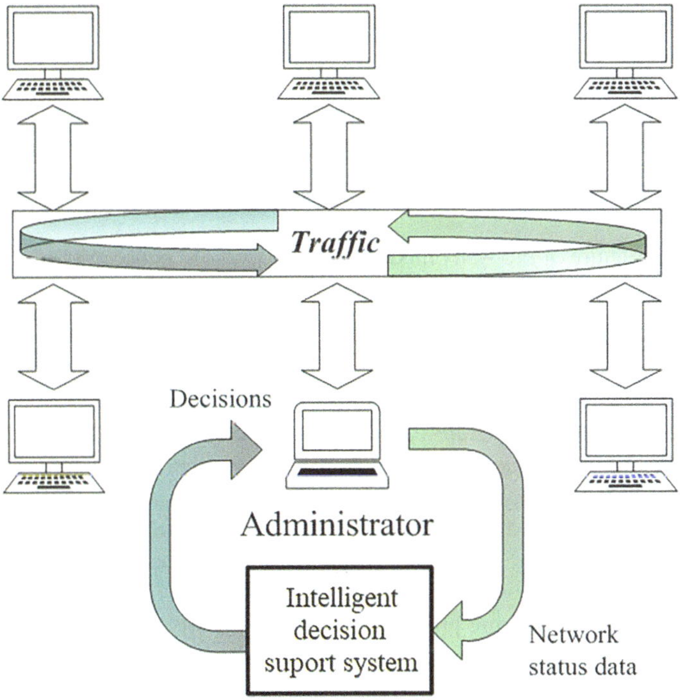

Fig. 1 The place of the intelligent decision support system in a network

rule, unidirectional work associated with the collection of data on the state of network elements. The disadvantage of this location variant is the significant time and computational resources required for training neural networks. In addition, the disadvantages of the system include the impossibility of using it in computer networks with complex topology.

The second location can be a network administrator's workstation. In this case, the system not only collects data on the state of the IoT network elements but also develops solutions. At the administrator's workplace, the system must be installed in a trained form. This option for localizing the system will be considered further.

The algorithm for the functioning of the intelligent decision support system is specified below.

The functioning of the system begins with the collection of data on the state of the elements of a computer network. This data is collected at each of the nodes of the computer network at a specified time interval. Data on the state of the elements of a computer network in the considered formulation of the problem play the role of initial data.

Then the data on the state of IoT network elements is transferred to the administrator's computer, where it is subjected to generalization and neural network processing. Neural network processing involves two stages: training and testing. After training the neural networks and bringing the intelligent decision support system into operation, the polling of the nodes of the computer network continues at a predetermined time interval. The resulting arrays of information after the ANN testing are used by an intelligent decision support system for the administrator to develop solutions.

The analytical block is the core of the proposed intelligent system for administering support. The analytical block is based on the combined ANN module. For the creation and functioning of the analytical unit of the intelligent decision support system, it is required:

- Set the structure of the ANN module.
- Determine the relationship between the objects of the ANN module.
- Select a set of indicators that most fully characterize all monitored aspects of the computer network functioning.

Then it is necessary to choose a method for normalizing the input data and determine the internal structure of the artificial neural network and the sequence of experiments. At the end of the solution of the problem, it is necessary to draw conclusions based on the results obtained during the experiments.

The collected data, which is fed to the input of the intelligent decision support system of the network administrator, is a set of various indicators characterizing the state of the computer network elements at different times. These indicators are used to characterize the following properties of the IoT network: transparency, network performance, reliability, cybersecurity, and throughput.

The sources of the collected information can be:

- System logs of network devices

- Network management and control systems based on the SNMP protocol

It should be noted that the following requirements are imposed on the indicators:

- Ease of obtaining
- The ability to obtain complete, redundant characteristics of a computer network
- Possibility of digital representation

The fulfillment of these requirements is necessary, since this greatly simplifies the learning process of the ANN and, accordingly, simplifies the configuration of the analytical unit of the intelligent decision support system.

Table 1 presents the proposed set of basic indicators, the values of which can characterize the state of the IoT network.

Table 1 IoT network status indicators

Property	Indicator	Formula	Comments
Transparency	Number of data per error	$Q = W \cdot \mu$	W – amount of data processed
			μ – average number of errors during data processing
Network performance	Average packet delay	$M = \sum_{i=1}^{N} T_i$	N – number of transmitted packets
			T_i – transmission time of i-th packet
	Response time to user request	$T_a = T_{at} + T_{ra} +$	T_{at} – request preparation time
		$+T_{an} + T_p + T_{pa}$	T_{ra} – transmission time for a request to the service server
			T_{an} – response preparation time
			T_p – response time
			T_{pa} – response processing time
Reliability	Reliability	$T = 1/\lambda$	λ – error rate
Cybersecurity	Operative availability factor	$K = k_r P(t)$	k_r – availability factor
			$P(t)$ – probability of failure-free operation in time t
Throughput	Effective data rate	$V = N_b/T_s$	N_b – amount of useful information
			T_s – communication session duration

4 Architecture

4.1 General Architecture of the Analytical Unit

The general architecture of the analytical unit of the intelligent decision support system for IoT network administrators is shown in Fig. 2.

The architecture includes:

- **Data normalization module** designed to bring the initial data to the form necessary for the correct training and operation of artificial neural networks. If necessary, this module converts non-digital values of indicators into digital form.
- **Artificial neural networks module** necessary for processing the values of indicators characterizing key processes taking place in a computer network. It includes a complex of artificial neural networks, the structure of which will be discussed below.
- **Results analysis module** obtained is necessary for the correct presentation to the end user of the results of neural networks.

For the analytical unit to work correctly, it is necessary to provide in the intelligent support system an expandable and self-updating repository of knowledge on the main aspects of the functioning of a computer network. This repository should include the knowledge necessary to calculate indicators of the state of the network, knowledge about the network topology and its composition, etc.

4.2 Data Normalization Module

For data normalization, it is proposed to use methods of linear or nonlinear normalization.

Linear normalization uses the minimax principle. It is executed according to the following expression:

$$\widetilde{x_i} = \frac{x_i - x_{min}}{x_{max} - x_{min}}, \quad (1)$$

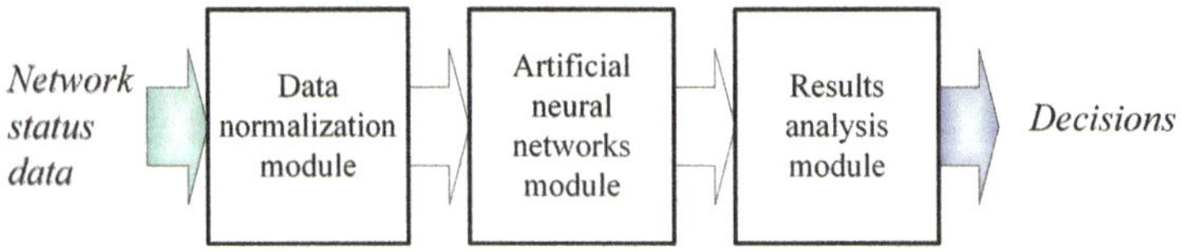

Fig. 2 The architecture of the analytical unit

where i is the interval number, x_i is a value for i-th interval, and x_{min} and x_{max} are the minimal and maximal values of the converted sequence, respectively.

For nonlinear normalization, a sigmoid logistic function is used, varying within [0, 1]:

$$\widetilde{x}_{ik} = \frac{1}{e^{-a(x_{ik}-x_{ck})} + 1}, \tag{2}$$

where x_{ik} and $\widetilde{x}_{ik}$ are the initial and normalized values of the k-th data element of the i-th interval, respectively, x_{ck} is the center of the normalized i-th interval, and a is the function slope parameter.

It is advisable to use normalization by a linear function for the values of indicators, the minimum and maximum values of which differ by no more than ten orders of magnitude. Otherwise, it is better to use nonlinear function normalization. This approach to the preparation of initial data can significantly reduce the time required for training artificial neural networks.

4.3 Artificial Neural Networks Module

The structure of the ANN module is shown in Fig. 3. The module is based on a combined neural network that includes a Kohonen layer (self-organizing map, SOM), one or more blocks, where each block consists of three layers of neurons (with linear, nonlinear, and RELU activation functions). The linear activation function layer is also used as the output layer of the neural network.

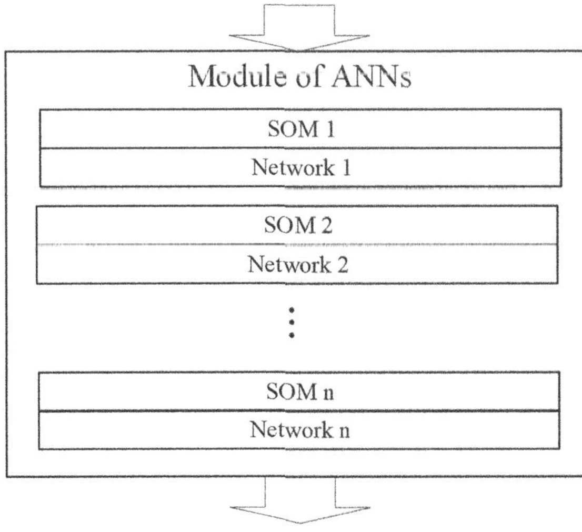

Fig. 3 The structure of the module of ANNs

Fig. 4 The Kohonen layer (self-organizing map)

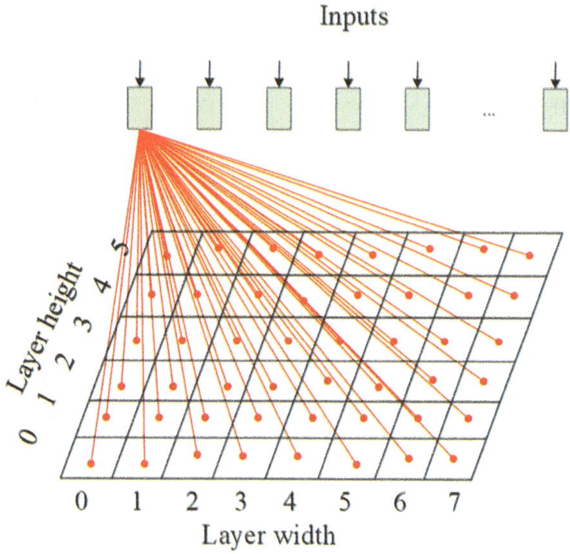

Self-organizing maps analyze the location of the points of the input layer and reproduce the metric proximity of vectors through the output layer (Fig. 4). Thus, they determine the degree of regularity of the source data.

The functioning of self-organizing maps is one of the options for clustering multidimensional vectors. An example of such algorithms is the k-nearest average algorithm. An important difference between a self-organizing map is that all neurons in it are ordered into a certain structure. Typically, this structure is a two-dimensional grid (as shown in Fig. 4). During training, not only the winning neuron is modified but also its neighbors but to a lesser extent. Due to this, a self-organizing map can be considered one of the methods for reflecting a multidimensional space onto a space with a lower dimension. When using this type of neural network, vectors that are similar in the original multidimensional space turn out to be nearby and on the resulting two-dimensional map.

Before starting SOM training, you need to initialize the weights of the neurons. A well-chosen initialization method can significantly speed up learning and lead to better results. There are three ways to initiate initial weights:

- Initialization by random values (all weights are given small random values).
- Initialization by examples (the values of randomly selected examples from the training sample are set as initial values).
- Linear initialization (weights are initiated by the values of vectors linearly ordered along a linear subspace between the two principal eigenvectors of the original dataset).

Eigenvectors can be found, for example, using the Gram-Schmidt procedure [30]. However, in our work, we used random initialization as the simplest one.

The essence of SOM training is to carry out sequential modification of neurons. Each neuron is an n-dimensional vector $W = [w_1, w_2, \ldots, w_n]^T$, where W_i is a weight coefficient and n is determined by the dimension of the original space. At each step of SOM training, one of the vectors is randomly selected from the original dataset. Then a search is made for the most similar vector to it. From the found vectors, a winning neuron is selected that most closely resembles the input vector. The measure of similarity is the distance between vectors, calculated in Euclidean space. If we designate the winning neuron as c, then the criterion for its selection can be written as follows:

$$\|\widetilde{x} - W_c\| = min_k \|\widetilde{x} - W_k\|, \tag{3}$$

where $\widetilde{x}$ is the input data vector and W_k is the neuron-neighbor in relation to the input data vector.

To regulate the weight tuning procedure, we chose the competitive Hebbian Learning algorithm, which is considered the most acceptable for SOM [11]. This algorithm takes into account not only the contribution of the winning neuron but also the contribution of the nearest neighbors located in a certain radius. The algorithm contains the following steps.

Step 1. Initialization of weights w_i^0, $i = 1, 2, \ldots, n$, using random values within the boundaries $[-1; 1]$.

Step 2. The sequence of data images on which SOM will be trained is determined randomly.

Step 3. The inputs of a neural network iteratively apply the values of input data images. At each iteration, the distance $r_i = \rho(x, w_i)$ between the input data vector x and the weight vector w_i is calculated.

Step 4. The output neuron with the minimum distance and its closest neighbors are determined.

Step 5. For all neurons of the output layer, weights $\Delta\widetilde{w}_i = (\widetilde{w}_i - \widetilde{x})h(u, c, t)\gamma$ are calculated, where c is the index of the winning neuron, i is the index of the neuron with the weight vector $\Delta\widetilde{w}_i$, $\widetilde{x}$ is the input data vector, h is the environment function for the i-th neuron at the iteration t, and γ is the coefficient that determines the learning speed of the neural network.

Step 6. Weights correction: $W = W - \Delta W$.

Step 7. The next image is input. Learning continues iteratively until the stop criterion is reached.

In general, SOM performs the function of decreasing the dimension of the input data. Consequently, the training time for the neural network as a whole is reduced.

Let us now consider the structure of multilayer blocks in the ANN module. Each block (Fig. 5) includes three layers with the following activation functions:

- Linear $y = cx$, where c is a constant.
- Hyperbolic tangent expressed through the exponent $y = \frac{2^{2x}-1}{2^{2x}+1}$.
- RELU—the activation function is presented as $y = max(0; x)$.

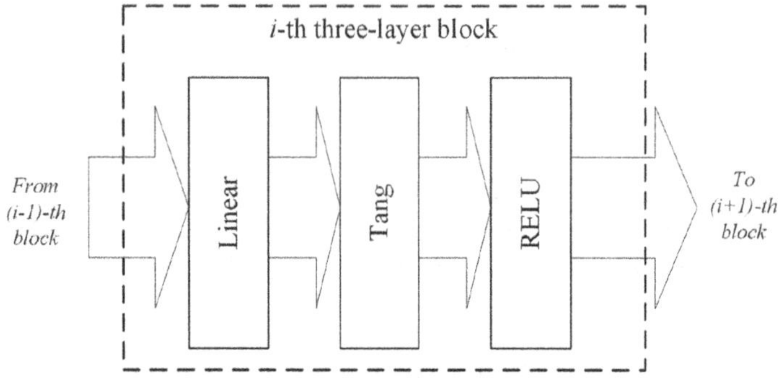

Fig. 5 The three-layer block

Multilayer blocks perform the function of summarizing and filtering incoming information. Their number and the number of active neurons depend on the dimension of the initial data received for processing.

The idea to use layers with different activation functions in the ANN module is based on the following assumptions. Each activation function has its own advantages and disadvantages. So, for a linear function, one of the easiest to implement, the disadvantage is the lack of sense in building multilayer structures, since several linear layers can always be successfully replaced with one linear layer. Nonlinear activation provides high accuracy in border areas as well as at high weights. The combined use of layers with these two types of activation function leads to the fact that the disadvantages inherent in each of the layers are eliminated and the advantages are preserved. As a result, the joint arrangement of these two layers makes it possible to reduce the number of neurons in both layers compared to using only a layer with a linear or nonlinear activation function. As a result, the network training time is reduced.

The third activation function combines the properties of linearity and nonlinearity. At all points except zero, it is nonlinear. At zero, it becomes nonlinear. Due to its linear nature, a layer with a RELU function can be used as a good approximator. Thus, the ANN becomes more robust with respect to the noise of the input dataset. In addition, the fact that the RELU function is zero in the negative region contributes to the discharge of the neural function in the ANN, i.e., many neurons in this layer acquire zero values. And this also provides a reduction in training time in the case of a very large dimension of the input vector.

The output line layer is necessary to reduce the dimension of the output data to the required level. It uses the linear activation function as the simplest.

The combined ANN test methodology is the typical one that is commonly used in data analysis problems using ANN. It includes the following stages: forming the initial data set, normalizing weight values, reducing the sample size by removing records with incomplete values from it, sample processing using ANN, and interpreting of the results obtained in accordance with the information available in the knowledge store.

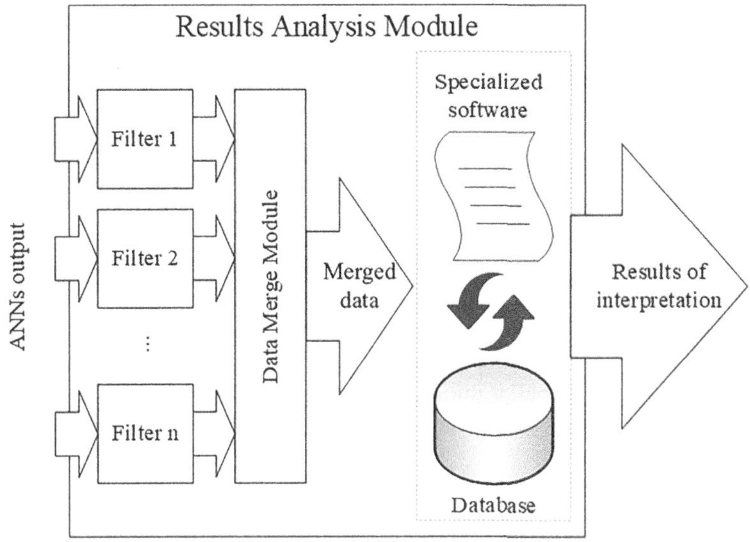

Fig. 6 The three-layer block

4.4 Results Analysis Module

The structure of the analysis block is shown in Fig. 6.

The block includes:

- Software filters that perform the function of rounding to thousandths after the decimal point, obtained at the outputs of artificial neural networks.
- Software module required to combine the values of the outputs of artificial neural networks into a single averaged sequence.
- Database designed to store the reference values of the outputs of artificial neural networks and expert opinions corresponding to these values.
- Specialized software interacting with the database is necessary to interpret the values obtained at the output of the value combining module.

Information from the output of the results analysis unit is transmitted to the information display device of the IoT network administrator.

5 Implementation, Experiments, and Discussion

5.1 Implementation

Python was used as a programming language for the implementation of the analytical block. Artificial neural networks were created using the PyTorch library. The computer on which the experiments were carried out had the following characteristics: an Intel Core i5 processor, DDR4 16Gb RAM, a GeForce RTX 3060 video card, and a 128 GB SSD.

ANN training is carried out in accordance with the algorithm described below (Fig. 7). The learning process ends when the minimum error threshold on the test sample is reached.

For training the ANNs that are part of the analytical unit, ready-made datasets were also available for download on the official website of the Australian Defense Forces Academy [1]. The dataset is designed to train the neural networks that make up host intrusion detection systems running on Linux and Windows operating systems. The performance of artificial neural networks was tested in a series of experiments for various configurations and sets of initial data. The effectiveness of various training methods for these configurations was also evaluated. Testing was carried out on a sample of initial data, including 10,000 objects. The sequence of data input to the neural network input was determined randomly. 150 training iterations were performed.

5.2 *Experiments on Training Error Evaluation*

The results of estimating the learning error for different dimensions and for different modes of using SOM (using SOM and without using SOM) are presented in Table 2. This table considers the following dimensions of the input data: 20 × 20; 50 × 50; 100 × 100; 150 × 150; 200 × 200; 250 × 250 and 300 × 300.

The structure of the ANN module in the "Without SOW" mode is shown in Fig. 8. Figure 9 shows the structure of the ANN when the SOM layer is used.

The experimental results obtained on the estimation of the training error fully confirmed the correctness of our hypothesis that the SOM layer provides a reduction in the dimension of the input data and reduces the training error. For all the dimensions of the network considered in the experiments, the sufficient number of neurons in the hidden layers in the "Without SOM" mode is 35–80 % higher than the same number for the "Without SOM" mode. The gain in the number of neurons in the hidden layers was ensured by the inclusion of SOM in the ANN module. And the fewer the number of neurons in the hidden layer, the less time it takes to train the network.

In general, the dependences of training error on time (time is expressed by the number of training epochs) for modes "With SOM" and "Without SOM" for dimensions 20 × 20, 50 × 50, 100 × 100, and 200 × 200 are shown in Figs. 10 and 11.

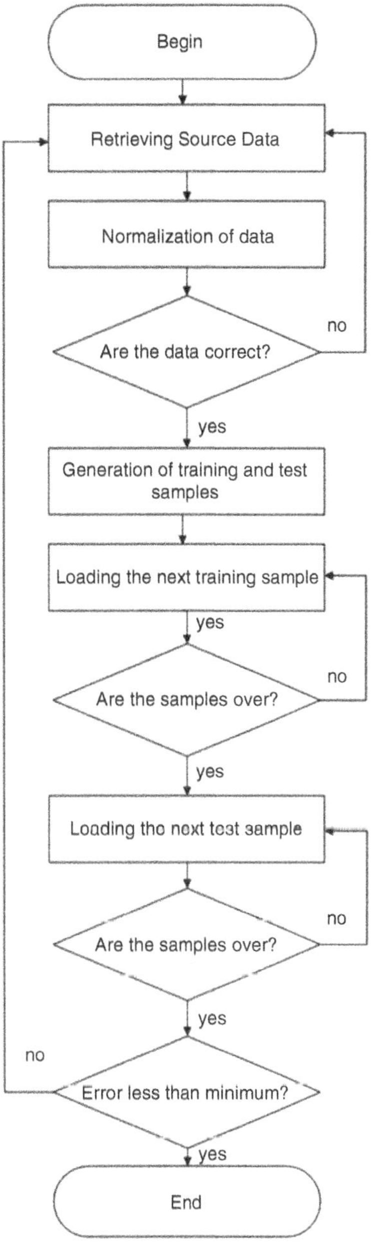

Fig. 7 ANNs training algorithm

Table 2 The minimum training error of a combined neural network without reducing the dimension of the input data

Input dimension	Number of multilayer blocks	The number of neurons in the hidden layer		Training error			
				With SOM		Without SOM	
		With SOM	Without SOM	20 epochs	Minimal	20 epochs	Minimal
20 × 20	2	100	180	0.0011	0.0011	0.0012	0.0012
50 × 50	3	250	450	0.0016	0.0016	0.006	0.0018
100 × 100	3	520	800	0.004	0.0015	0.011	0.0018
150 × 150	3	780	1200	0.008	0.0016	0.013	0.0019
200 × 200	4	1100	1600	0.012	0.0018	0.015	0.0021
250 × 250	4	1450	2000	0.016	0.0019	0.018	0.0023
300 × 300	4	1800	2500	0.020	0.0021	0.022	0.0025

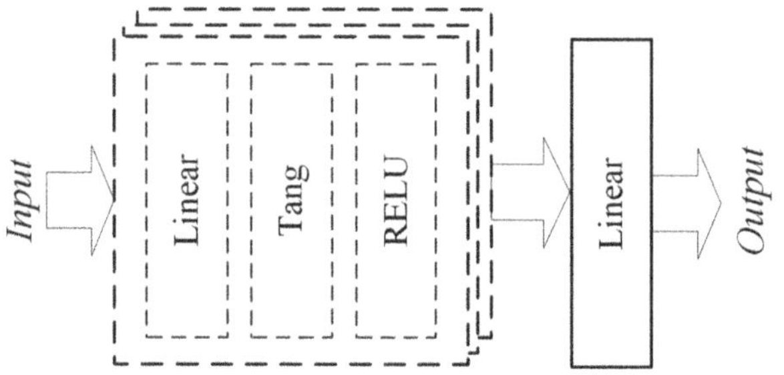

Fig. 8 ANN configuration for an analytic unit without a Kohonen layer

Fig. 9 ANN configuration for an analytic unit with a Kohonen layer

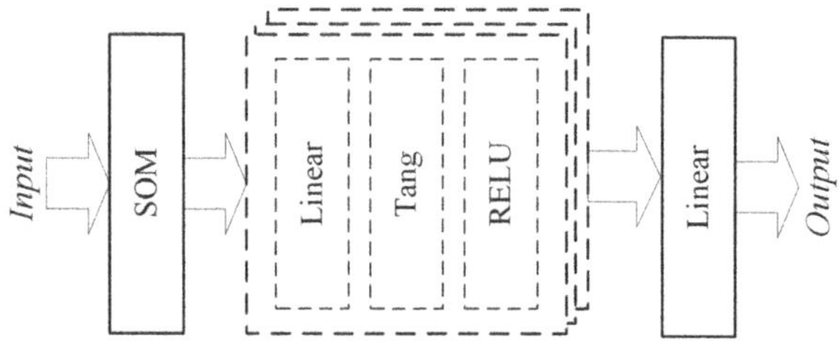

5.3 *Experiments on Compatible Evaluation of Training Methods*

In the course of the experiments, a comparative analysis of the effectiveness of various training methods was carried out in relation to the architecture of an artificial neural network discussed above. The following methods were compared:

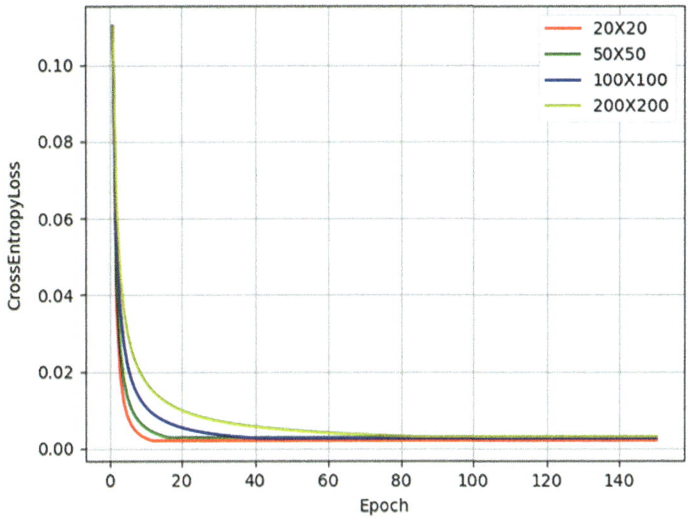

Fig. 10 Dependencies of ANN learning error on time (number of epochs) with SOM

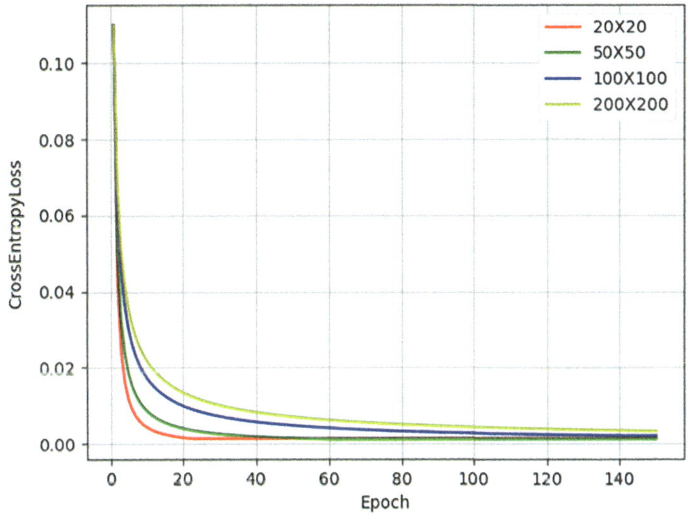

Fig. 11 Dependencies of ANN learning error on time (number of epochs) without SOM

- Stochastic gradient descent.
- The adaptive learning step method
- The adaptive inertia method.

Stochastic gradient descent (SGD) was considered as the first comparable method [3]. This method updates each of the parameters by subtracting the gradient of the optimized function for the corresponding parameter. In this case, the learning step η is scaled:

$$\theta_{t+1} = \theta_t - \eta \cdot grad(f_i(\theta_i)), \tag{4}$$

where f_i is a function calculated on the i-th part of the data, $grad(f_i(\theta_i))$ is a gradient of the function f_i, and t is an iteration step.

The adaptive learning step method (Adadelta) [44] uses an exponential moving average to estimate the second moment of the gradient $grad(f_i(\theta_i))$. The parameters are updated according to the following formula:

$$g_{t+1} = \gamma g_t + (1 - \gamma) \cdot grad(f_i(\theta_i))^2, \tag{5}$$

where γ is the hyperparameter, $grad(f_i(\theta_i))^2$ is the second moment of the gradient of the function f_i, and g_t is the estimation of the second moment of the gradient at the moment t.

The scaling parameter ν for $(t + 1)$ is calculated as

$$\nu_{t+1} = \frac{\sqrt{x_t + \varepsilon} \cdot grad(f_i(\theta_i))}{\sqrt{g_{t+1} + \varepsilon}}, \tag{6}$$

where ε is the constant introduced for numerical stability and x_t is the moving average at the moment t.

The moving average for $(t + 1)$ is calculated as

$$x_{t+1} = \gamma x_t + (1 - \gamma) \cdot \nu_{t+1}, \tag{7}$$

The final formula for updating each of the parameters for the Adadelta method is as follows:

$$\theta_{t+1} = \theta_t + \nu_{t+1}, \tag{8}$$

To implement the adaptive inertia (Adam) method, estimates of the first and second moments are initialized with zeros followed by a small correction. Then the estimate of the first moment is calculated as a moving average:

$$m_{t+1} = \gamma_1 m_t + (1 - \gamma_1) grad(f_i(\theta_i)), \tag{9}$$

$$g_{t+1} = \gamma_2 g_t + (1 - \gamma_2) grad(f_i(\theta_i))^2, \tag{10}$$

where m is the first moment assessment, g is the second moment assessment, and γ_1 and γ_2 are hyperparameters.

Correction of assessments for the moment $(t + 1)$ is carried out according to following formulas:

Table 3 Average training error for various training methods

Learning method	Training error after 20 epochs	Minimal training error
SGD	0.011	0.0034
Adadelta	0.0015	0.0014
Adam	0.0025	0.0021

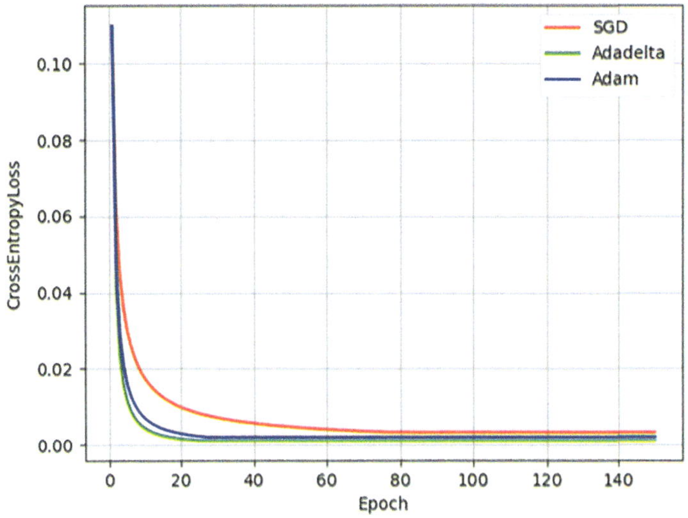

Fig. 12 The dependence of the error of the output of the neural network on the training time

$$\widehat{m}_{t+1} = \frac{m_{t+1}}{1 - \gamma_1^{t+1}} \tag{11}$$

$$\widehat{g}_{t+1} = \frac{g_{t+1}}{1 - \gamma_2^{t+1}} \tag{12}$$

Parameters are updated as follows:

$$\theta_{t+1} = \theta_t - \frac{\eta \widehat{m}_{t+1}}{\widehat{g}_{t+1} + \varepsilon} \tag{13}$$

where η is the hyperparameter and ε is the constant introduced for numerical stability.

The training outcomes of ANNs are presented in Table 3.

The dependence of the error of the output of the neural network on the training time (the number of iterations of training) is presented in Fig. 12.

The found dependences show that with a short training time (i.e., 20 epochs), there is no big difference between the Adam and Adadelta methods. At the same time, the SGD method has a much larger error (it is 5–10 times larger).

Table 4 Test results of the analytical unit

Dataset size, 10^3	Number of multilayer blocks	False interpretation of the result, %	Testing time, sec
15	2	37	3.13
	3	20	4.23
	4	12	5.52
25	2	19	5.35
	3	12	7.18
	4	9	9.25
30	2	14	6.42
	3	9	8.43
	4	6	11.55

With a long training time, all methods have approximately the same and rather small error. However, the best method is the Adadelta method.

5.4 Experiments on Combined ANN Testing

The trained combined ANNs were tested on datasets of various sizes, containing 15,000, 25,000, and 30,000 records. At the same time, the number of multilayer blocks in the combined ANN changed: 2, 3, and 4.

The result of the assessment was the value of the "false interpretation of the result" indicator, which is defined as the sum of the false positive rate (FPR) and false negative rate (FNR) indicators. In turn, FPR and FNR are defined as follows:

$$FPR = FP/N, FNR = FN/N, \tag{14}$$

where N is the size of the testing sample, $N = TP + TN + FP + FN$, TP is the number of correct positive decisions, TN is the number of correct negative decisions, FP is the number of erroneous positive decisions, and FN is the number of erroneous negative decisions.

The test results of the analytical unit are shown in Table 4.

Analyzing the data presented in Table 4, the following conclusions can be drawn.

The best result in terms of the accuracy of interpreting the network state was shown by the analytical unit, which has 4 multilayer units that were trained on a dataset of 30,000 records. The accuracy of interpretation decreases with a decrease in the number of multilayer blocks and a decrease in the volume of the tested dataset. Moreover, this dependence is close to linear.

Analyzing the data on the testing time, we can conclude that the testing time depends linearly on the volume of the dataset and on the number of multilayer blocks used. So, if the volume of the dataset is doubled (from 15,000 to 30,000), then the testing time will also increase approximately 2 times.

If the number of multilayer blocks is doubled, then the testing time will also increase by about two times. This effect is explained by how many neurons are present in the neural network and how many times they will be used during testing. The greater the number of multilayer blocks, the greater the number of neurons in the network. On the other hand, the larger the number of records in the dataset, the more times neurons will be used during ANN testing. These factors affect testing time.

5.5 *Common Discussion*

The experimental results show a tendency for the neural network error to increase with an increase in the dimension of the input data sample. In this regard, in order to minimize the error, it is necessary to increase the number of neurons in the hidden layers or the number of hidden layers.

The use of the Kohonen layer as an input filter allows one to reduce the size of the neural network and the required training time with a slight increase in the error at its outputs.

The most effective of the considered training methods turned out to be the adaptive learning step method (Adadelta), which showed the minimum error with the shortest network training time.

The classification error when testing trained neural networks does not exceed 6.4% when using input data sets obtained by the formulas presented in Table 1.

When using ready-made data sets for training neural networks [1], the classification error during testing was no more than 16.7%.

In [37], a similar dataset was used to assess the anomalous patterns that appear in the normal pattern and the anomalous pattern of the dataset using data mining techniques such as support vector classification and random forest classification. The classification error obtained in this article is 33–46

The results obtained in the course of the study clearly demonstrate the rather high efficiency of the proposed architecture of artificial neural networks as part of the analytical unit of the intelligent support system for the IoT network administrator.

6 Conclusion

The chapter proposed an approach to the implementation of an intelligent decision support system and an analytical unit based on ANN for an intelligent decision support system for IoT network administrator. The proposed analytical unit includes three modules: input data normalization module, ANN module, and results analysis module.

The architecture of the ANN module represents a set of sequentially modified layers that are connected in series as follows: SOM block, two or more three-layer neural blocks, and output linear layer. Each of the three-layer neural blocks is a combination of series-connected neural layers with a linear, nonlinear, and RELU neuron activation function. The proposed ANN module architecture is flexible and can be adapted to various input configurations.

The SOM block in the proposed architecture of the combined ANN is used as a filter that allows one to reduce the dimension of the input data and, thereby, reduce the training time. In addition, SOM made it possible to simplify the ANN structure without significantly reducing the quality of the calculation results.

Experimental estimates have shown that among the known training methods, the Adadelta method is most suitable, for which the classification error when testing trained neural networks did not exceed 6.4%.

Further research directions are supposed to be carried out in the field of application of other machine learning mechanisms for solving network administration problems, such as support vector machine, decision tree, kNN, etc., [6], as well as combining these approaches with security event ontologies developed for security information and event management systems [19], and using cooperative defense mechanisms [17, 18, 22].

Acknowledgments This research is being supported by the grant of RSF #21-71-20078 in SPC RAS.

References

1. ADFA. IDS dayasets of the Australian Defense Force Academy. https://www.unsw.adfa.edu.au/unsw-canberra-cyber/cybersecurity/ADFA-IDS-Datasets, 2021. Online: accessed 05.05.2020.
2. G. A. Akpakwu, B. J. Silva, G. P. Hancke, and A. M. Abu-Mahfouz. A survey on 5g networks for the internet of things: Communication technologies and challenges. *IEEE access*, 6:3619–3647, 2017.
3. S. Amari. A theory of adaptive pattern classifiers. *IEEE Transactions on Electronic Computers*, (3):299–307, 1967.
4. A. Azruddin, R. Gobithasan, B. Rahmat, S. Azman, and R. Sureswaran. A hybrid rule based fuzzy-neural expert system for passive network monitoring. In *Proc. of the Arab Conf. on Information Technology ACIT*, pages 746–752, 2002.
5. S. Behal, K. Kumar, and M. Sachdeva. D-face: An anomaly based distributed approach for early detection of ddos attacks and flash events. *Journal of Network and Computer Applications*, 111:49–63, 2018.
6. A. Branitskiy and I. Kotenko. Hybridization of computational intelligence methods for attack detection in computer networks. *Journal of Computational Science*, 23:145–156, 2017.
7. A. A. Brincat, F. Pacifici, S. Martinaglia, and F. Mazzola. The internet of things for intelligent transportation systems in real smart cities scenarios. In *2019 IEEE 5th World Forum on Internet of Things (WF-IoT)*, pages 128–132. IEEE, 2019.
8. D. Bukhanov and V. Polyakov. An approach to improve the architecture of art-2 artificial neural network based on multi-level memory. *Fuzzy Technol. Ind. FTI*, 2018:235–242, 2018.

9. Y. Chen, S. Kak, and L. Wang. Hybrid neural network architecture for on-line learning. *Intelligent Information Management*, page 253, 2010.
10. O. Friha, M. A. Ferrag, L. Shu, L. A. Maglaras, and X. Wang. Internet of things for the future of smart agriculture: A comprehensive survey of emerging technologies. *IEEE CAA J. Autom. Sinica*, 8(4):718–752, 2021.
11. M. Ghofrani, R. Azimi, F. Najafabadi, and N. Myers. A new day-ahead hourly electricity price forecasting framework. In *2017 North American Power Symposium (NAPS)*, pages 1–6. IEEE, 2017.
12. W. Iqbal, H. Abbas, M. Daneshmand, B. Rauf, and Y. A. Bangash. An in-depth analysis of iot security requirements, challenges, and their countermeasures via software-defined security. *IEEE Internet of Things Journal*, 7(10):10250–10276, 2020.
13. K. Ishida. Iot application in sports to support skill acquisition and improvement. In *2019 IEEE 12th Conference on Service-Oriented Computing and Applications (SOCA)*, pages 184–189. IEEE, 2019.
14. A. Karmakar, N. Dey, T. Baral, M. Chowdhury, and M. Rehan. Industrial internet of things: a review. In *2019 international conference on opto-electronics and applied optics (optronix)*, pages 1–6. IEEE, 2019.
15. N. Kasabov and H. N. Hamed. Quantum-inspired particle swarm optimisation for integrated feature and parameter optimisation of evolving spiking neural networks. *International Journal of Artificial Intelligence*, 7(A11):114–124, 2011.
16. O. Khristodulo, A. Makhmutov, and T. Sazonova. Use algorithm based at hamming neural network method for natural objects classification. *Procedia Computer Science*, 103:388–395, 2017.
17. I. Kotenko. Multi-agent modelling and simulation of cyber-attacks and cyber-defense for homeland security. In *2007 4th IEEE Workshop on Intelligent Data Acquisition and Advanced Computing Systems: Technology and Applications*, pages 614–619. IEEE, 2007.
18. I. Kotenko, A. Konovalov, and A. Shorov. Agent-based simulation of cooperative defence against botnets. *Concurrency and Computation: Practice and Experience*, 24(6):573–588, 2012.
19. I. Kotenko, O. Polubelova, and I. Saenko. The ontological approach for siem data repository implementation. In *2012 IEEE International Conference on Green Computing and Communications*, pages 761–766. IEEE, 2012.
20. I. Kotenko, I. Saenko, and F. Skorik. Intelligent support for network administrator decisions based on combined neural networks. In *13th International Conference on Security of Information and Networks*, pages 1–8, 2020.
21. I. Kotenko, I. Saenko, F. Skorik, and S. Bushuev. Neural network approach to forecast the state of the internet of things elements. In *2015 XVIII international conference on soft computing and measurements (SCM)*, pages 133–135. IEEE, 2015.
22. I. Kotenko and A. Ulanov. Simulation of internet ddos attacks and defense. In *International Conference on Information Security*, pages 327–342. Springer, 2006.
23. S. D. Kumar et al. Design and development of iot-based robot. In *2020 International Conference for Emerging Technology (INCET)*, pages 1–4. IEEE, 2020.
24. I. Kurochkina, I. Kalinin, L. Mamatova, and E. Shuvalova. Neural networks method in modeling of the financial company's performance. *Statistics and Economics*, (5):33–41, 2017.
25. C. K. Lee, C. L. Yeung, and M. N. Cheng. Research on iot based cyber physical system for industrial big data analytics. In *2015 IEEE International Conference on Industrial Engineering and Engineering Management (IEEM)*, pages 1855–1859. IEEE, 2015.
26. A. Mishra, Z. Zaheeruddin, et al. Design of hybrid fuzzy neural network for function approximation. *Journal of Intelligent Learning Systems and Applications*, 2(02):97, 2010.
27. S. J. Moore, C. D. Nugent, S. Zhang, and I. Cleland. Iot reliability: a review leading to 5 key research directions. *CCF Transactions on Pervasive Computing and Interaction*, 2(3):147–163, 2020.
28. C. Perera, A. Zaslavsky, P. Christen, and D. Georgakopoulos. Context aware computing for the internet of things: A survey. *IEEE communications surveys & tutorials*, 16(1):414–454, 2013.

29. E. Popova and V. Leonenko. Predicting user reactions in social networks using machine learning methods. *Bulletin of Information Technologies, Mechanics and Optics*, 20(1):118–124, 2020.
30. S. M. Rao. Application of modified gram-schmidt procedure to obtain induced currents on a section of a large body. In *2014 IEEE Antennas and Propagation Society International Symposium (APSURSI)*, pages 2008–2009. IEEE, 2014.
31. A. K. Ray and A. Bagwari. Iot based smart home: Security aspects and security architecture. In *2020 IEEE 9th international conference on communication systems and network technologies (CSNT)*, pages 218–222. IEEE, 2020.
32. S. Rizvi, A. Kurtz, J. Pfeffer, and M. Rizvi. Securing the internet of things (iot): A security taxonomy for IoT. In *2018 17th IEEE International Conference On Trust, Security And Privacy In Computing And Communications/12th IEEE International Conference On Big Data Science And Engineering (TrustCom/BigDataSE)*, pages 163–168. IEEE, 2018.
33. I. Saenko, F. Skorik, and I. Kotenko. Application of hybrid neural networks for monitoring and forecasting computer networks states. In *International Symposium on Neural Networks*, pages 521–530. Springer, 2016.
34. M. U. Saleem, M. R. Usman, and M. Shakir. Design, implementation, and deployment of an iot based smart energy management system. *IEEE Access*, 9:59649–59664, 2021.
35. N. Shahid and S. Aneja. Internet of things: Vision, application areas and research challenges. In *2017 International Conference on I-SMAC (IoT in Social, Mobile, Analytics and Cloud)(I-SMAC)*, pages 583–587. IEEE, 2017.
36. L. C. Silva, E. F. Simas Filho, M. C. Albuquerque, I. C. Silva, and C. T. Farias. Embedded decision support system for ultrasound nondestructive evaluation based on extreme learning machines. *Computers & Electrical Engineering*, 90:106891, 2021.
37. C. K. Simon and I. V. Sochenkov. Evaluating host-based intrusion detection on the adfa-wd and ADFA-WD: SAA datasets. *Semanticscholar. org*, 2021.
38. S. Sinche, D. Raposo, N. Armando, A. Rodrigues, F. Boavida, V. Pereira, and J. S. Silva. A survey of iot management protocols and frameworks. *IEEE Communications Surveys & Tutorials*, 22(2):1168–1190, 2019.
39. S. Sinche, J. S. Silva, D. Raposo, A. Rodrigues, V. Pereira, and F. Boavida. Towards effective iot management. In *2018 IEEE SENSORS*, pages 1–4. IEEE, 2018.
40. L. G. M. Souza and G. A. Barreto. Nonlinear system identification using local arx models based on the self-organizing map. *Learning and nonlinear models-revista da sociedade brasileira de redes neurais (SBRN)*, 4(2):112–123, 2006.
41. S. Vishnu, S. J. Ramson, and R. Jegan. Internet of medical things (IoMT)-an overview. In *2020 5th international conference on devices, circuits and systems (ICDCS)*, pages 101–104. IEEE, 2020.
42. L. Wan, L. Zhu, and R. Fergus. A hybrid neural network-latent topic model. In *Artificial Intelligence and Statistics*, pages 1287–1294. PMLR, 2012.
43. J. Whitter-Jones. Security review on the internet of things. In *2018 Third International Conference on Fog and Mobile Edge Computing (FMEC)*, pages 163–168. IEEE, 2018.
44. M. D. Zeiler. Adadelta: an adaptive learning rate method. *arXiv preprint arXiv:1212.5701*, 2012.

A Novel Privacy-Preserving Framework Based on Blockchain Technology to Secure Industrial IoT Data

Aitizaz Ali, Mehwish Kundi, Mehmood Ahmed, Ateeq ur Rehman, Hashim Ali, and Aervina Binti Misron

1 Introduction

Blockchain is usually described as a distributed ledger which permanently stores the transactions that occur between the participants into blocks. The main characteristics of blockchain are:

1. Decentralised application without central authority
2. Fast reconciliation between parties compared to traditional central verifiers
3. Transparent, verifiable, and auditable technology that allows every connected asset to track the status of the ledger
4. Communication in a trustless environment

A. Ali (✉) · A. B. Misron
Faculty of Business and Technology, Unitar International University, Kelana Jaya, Malaysia
e-mail: aitizaz.ali@unitar.my; aervina.misron@unitar.my

M. Kundi · H. Ali
Department of Computer Science, Abdul Wali Khan University, Mardan, KPK, Pakistan
e-mail: istanmkundi@awkum.edu; hashimali@awkum.edu.pk

M. Ahmed
Department of Information Technology, University of Haripur, Haripur, KPK, Pakistan
e-mail: mehmood@uoh.edu.pk

A. ur Rehman
Department of Biomedical Engineering, Foundation University Islamabad, Islamabad, Pakistan
e-mail: ateequr.rehman@fui.edu.pk

L. Fotia et al. (eds.), *Security, Trust and Privacy Models, and Architectures in IoT Environments*, Internet of Things, https://doi.org/10.1007/978-3-031-21940-5_2

According to Gartner, blockchain is predicted to produce more than $3 trillion in annual corporate value by 2030. Internet of Things (IoT)-based solutions are becoming more interested in using blockchain technology because of these interesting studies and trends [1–4]. According to a Business Insider research, by 2027, the IoT market is expected to increase at an annual rate of more than 2.4 trillion dollars. The Industry 4.0 project indicated a great deal of interest in digitising industrial assets as a result of this rapid expansion. Artificial intelligence (AI), enhanced automation, and data analytics are all part of the fourth industrial revolution [3]. The increased connection within critical infrastructures, such as Industrial Internet of Things (IIoT), poses many security vulnerabilities as a result of this massive expansion. The use of Ethereum blockchain in IIoT contexts raises privacy concerns, which are the subject of this study [4, 5]. Because Ethereum is one of the most prominent blockchain ecosystems for creating decentralised applications (Dapps), it was chosen as the platform of choice. According to a report by *Coin Telegraph*, as of mid-2020, the number of Ethereum accounts outnumbers the number of Bitcoin addresses by around four to one. As a result of these issues, we set out to develop and execute a deep learning blockchain framework for boosting IIoT security and privacy. The following are some of the main points of this chapter. As a result of this study, the following results were attained:

- A new privacy-preservation and intrusion detection framework is designed by using blockchain and a deep learning technique that enables two-level privacy and two-level security in IIoT. Moreover, this framework presents a generalised way to transmit industrial data to the cloud by leveraging edge nodes [6].
- In the first level of data security and privacy preservation, a blockchain and smart contract scheme is suggested to enable immutable data exchange and to avoid data poisoning attacks [5–7].

The remaining chapter is organised into six sections. Section 2 provides an overview of Ethereum and its anatomy. Section 3 presents the existing risks and the past attacks for IoT infrastructures. Section 4 explains and discusses the existing privacy issues with the Ethereum blockchain and our research questions. Section 5 provides an in-depth security analysis of the permissions and rights allowed within an Ethereum network. Section 5 presents a threat model of a typical Ethereum blockchain topology and a risk analysis of the discussed threats following the risk management guidance produced by the National Institute of Standards and Technology (NIST) [8–10]. Section 5 provides an analysis of the explored threats with respect to an IIoT environment and explores the current protocols and frameworks that aim to improve the privacy aspects of Ethereum.

2 Background and Related Works

2.1 Background

In this section, we have discussed the main concepts that we have used in our proposed method. These concepts are blockchain, consensus protocol, and access control.

2.1.1 Blockchain Technology

These include consensus Protocol, Hashing, P2P topology, Immutable Ledger and mining [9]. In Fig. 2, the workings of a blockchain network, which includes various nodes and protocols, are clearly depicted.

Consortium blockchains are permissionless blockchains that have been adopted by a wide range of blockchain enthusiasts. As with all blockchain networks, the consortium blockchain adheres to the basic principles and components. The proof of work (PoW) consensus technique is used as a blockchain to allow mining nodes to validate network blocks before they are added to the blockchain and their transactions are recorded [10]. A unique chain-id can be used to link Ethereum nodes to a blockchain. Nodes can now engage in blockchain operations and access blocks or transactions, regardless of where they are located. In typical enterprise applications, Ethereum can also be implemented as a private blockchain to restrict access to its trusted assets and people [11, 12]. The following are the most critical aspects of the Ethereum blockchain:

1. User-created "smart contracts" are one of Ethereum's most distinguishing features. A smart contract is a computerized transaction mechanism that performs the provisions of a contract, as defined by Nick Szabo in 1994 [12]. By utilizing smart contracts, scripts recorded in the blockchain can now be executed directly without a third party intermediary [13–16].
2. Transactions: Ethereum transaction is a signed package of data that contains:
    (a) The signature of the sender
    (b) The recipient of the message
    (c) The amount of ether to be transferred
    (d) Data field (optional)

2.1.2 Consensus Protocol

A consensus protocol provides the approval of a specific transaction if 51% of the users approved it [14]. Keywords related to consensus protocol are defined as below:

1. Hash cryptography: A blockchain hash is also known as SHA256. Its size is 64 characters.
2. A hash algorithm consists of applications such as cryptography, deterministic approach, and robust computation.
3. Immutable ledger: The transaction inside the blockchain framework are immutable [15].

2.1.3 Access Control

Authentication is a term that encompasses both access control and authentication. Unauthorised and authorized users can be kept out of a system using access control. There are predetermined rules that it follows. These regulations vary depending on the organization, role, or traits of the person. This fast-developing area of computer technology focuses on defending networks and information from unauthenticated access, theft, and data misuse as well as most security breaches such as collusion assaults and phishing.

Industrial systems use the Internet of Things (IIoT) to connect to the cloud for data collection and analysis. In many ways, IIoT and ICS share many parallels; nevertheless, the tight limitations on ICS cannot be immediately applied to IIoT contexts [16, 17]. It's because cloud computing requires direct internet connectivity for IIoT devices.

To deal with this kind of scenario in a traditional ICS system, you'll need zoning and defence-in-depth strategies. Figure 1 shows how ICS settings differ from regular business systems in a number of ways [18, 19]:

1. Risk level: Significant impact on human lives and possibly the nation.
2. Performance Requirements: Real-time analysis is necessary since performance faults might have an impact on operations and hence be risky for the company.
3. In the event of a system breakdown, the system must be available at all times with redundant systems.
4. Safety: By identifying anomalous behavior and initiating warnings and safety procedures, the system is able to prevent harm. The Industrial Control System (ICS) has been the target of numerous attacks in the past, all of which used malware payloads to seize control of the system. This includes but is not limited to the following ICS cyberattacks:
5. Stuxnet Malware [19].
6. Multiple attacks on Ukraine Power Grid in 2015–2016 [20].
7. Ransomware attacks by NotPetya [21].
8. TRITON attack framework targeting safety instrumented system [22].

We can see that the data transmitted and received by conventional ICS systems is extremely sensitive. The ramifications of unauthorized access to data can be far-reaching for both businesses and entire countries. In order to adopt IIoT, some of these systems may need to be accessible directly to the cloud. Various IoT devices

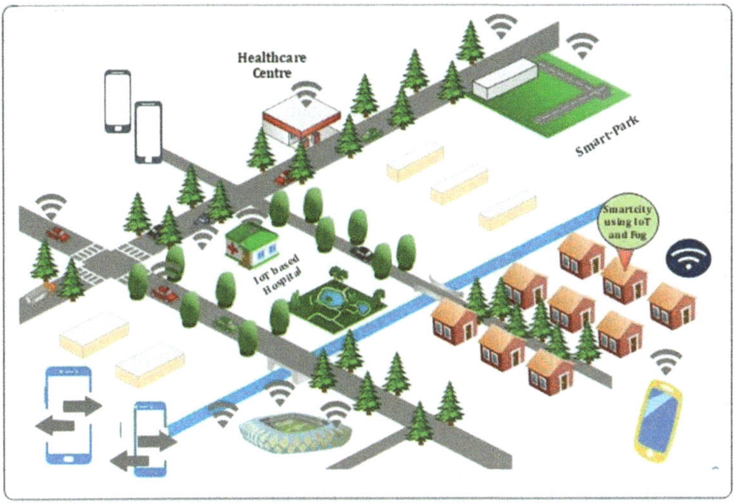

Fig. 1 Application of IoT in real life and healthcare systems

have been hacked to carry out Direct denial of Service (DDoS) attacks [23] or to send data to command-and-control (C2) servers.

2.2 *Related Works*

Data may be accessed in a decentralized manner, thanks to the blockchain. Federated learning has become the most popular method of using blockchain in the last few years. Our model is trained on local nodes, which is known as a local model, and then tested using data from all of the edge nodes through federated learning. As a result, less storage is required, latency is reduced, and computing costs are reduced. A proposal was presented in [22–25] to visualize and generate a graphical depiction of an Ethereum blockchain using graphtheoretic data mining techniques. The privacy of nodes has been a concern with this technology. This technique was employed by the authors in order to identify the most active and central addresses on the blockchain. In this way, the main IP address of the account might be targeted by an attacker [26]. In addition, scholars in [27] examined the loss of transaction privacy. Due to their entire transparency and the possibility of vulnerabilities that could be exploited, smart contracts are no exception [28]. It's important to note that while these hazards have been examined in general for blockchain, they haven't been examined in depth in relation to IIoT systems that are considered critical infrastructures. Table 1 summarizes the currently available research and the associated limitations. The following research problems were investigated in our article using proper threat modelling for a private Ethereum blockchain solution:

Table 1 Comparative analysis of attack resistance

Models	Collusion attacks	DoS	DDoS
Medblock	No	No	Yes
Casht et al.	Yes	No	No
Medchain	Yes	No	No
Kasra et al.	Yes	No	No
Proposed	Yes	Yes	Yes

1. What are the threats associated with IoT data?
2. How can blockchain transparency impact the exposure of IIoT environments to external threats?
3. What is the impact of compromising blockchain nodes within IIoT model?

3 Methodology

The IIoT model is based on the distribution of small sensing nodes in an IoT environment depending on the structure and size of the company. Cluster heads or base stations receive the data from the sensors nodes, which include the patient's body, heart rate, pulse rate, insulin level, and temperature. A middle man, denial of service attack, or a collusion attack can all compromise this data as it travels via the air. The integration of consortium blockchain is employed in the suggested approach to give assurance and confidence. Public and private interfaces are provided by the consortium blockchain. Private data is stored on the consortium blockchain in the consortium's private interface, while public data is displayed to the general public on the consortium blockchain. A private consortium blockchain topology with peer-to-peer nodes was created to test the interactivity and adaptability of a consortium blockchain system. 2. We generated four virtual machine (VM) and chain code was put on each VM and operated on Ubuntu OS, which is the light of Ubuntu Linux distribution with 1 vCPU and 2GB. In addition, we tracked the graphical processing unit's (GPU) performance, as well as the generation, validation, and encryption/decryption times for each block. For performance evaluation we observed delay rate, execution time and number of iterations. Figure 2 represents the flow of the transactions through blockchain and its structure.

3.1 Intrusion Detection and Its Protection

Security breaches and data breaches caused by insider threats have a significant impact on the likelihood of an internal threat. Insider assaults have occurred in 60% of the firms evaluated in 2019, according to a Nucleus Cyber analysis from that year. In addition, according to a report by Proof point [20], criminal or malevolent

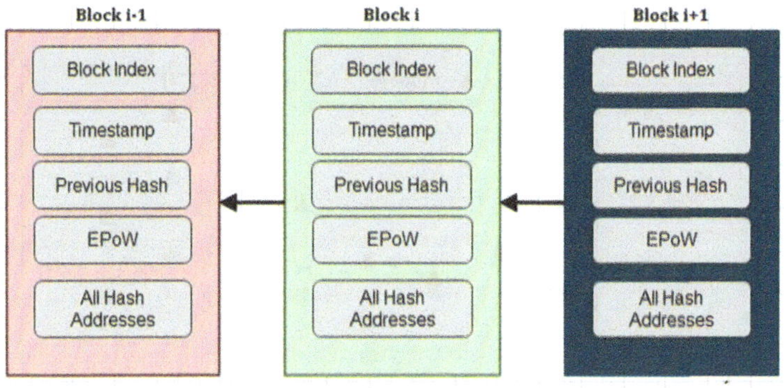

Fig. 2 Flow of transactions through blockchain and its structure

insiders will be responsible for 23% of recorded attacks in 2020. There is a reasonable chance that threat T1 may materialize [21]. On the basis of the frequency of security breaches as well as the complexity of assaults and their disclosure, external threats are more likely to occur [22]. According to Verizon's 2021 Data Breach Investigation Report [23], the following are the results of an external hostile actor-caused data breach. Phishing accounts for about 40% of data breaches, while malware accounts for the other 20%. As a result, the chance of the danger T5 is moderate. Hacking was a factor in almost 40 of the data breaches. According to threat T2's moderate likelihood, this applies [24]. Around 20 data breaches were caused due to error and around 20% were caused to error. This applies to threats T3 and T4, which makes their likelihood moderate.

The Internet of Things (ToN-IoT) dataset [25] can be used to assess the method's efficacy. The UNSW Canberra Cyber Range Center used a realistic network model to create the dataset. Message Queuing Telemetry Transport Protocol is the data collection protocol. DDoS, Injection, Password XSS, Backdoor, Scanning DoS, MITM (Man in the Middle), and Ransomware are all included in this collection. We used 70% of the dataset for training and 30% for testing for this experiment. The analysis of previously recorded connected network data is the second major phase in the SG inspection process. We tested the blockchain network in the simulation with various analysts operating at the same time. Every analysis must analyse Nominalize inputs.

4 Results

We've covered all of the findings from the simulations in detail in this section. Elapsed time or intervened time is the corresponding throughput while creating an ordering service at a Fog node (FN) for the Internet of Medical Things (IoMT) network. When a virtual machine (VM) is used to create a cloud-based ordering system, the bypass time and the number of nodes are measured. Moreover, during the experiment Core i5-CPU 16.04 LTS is used for performance testing. Each

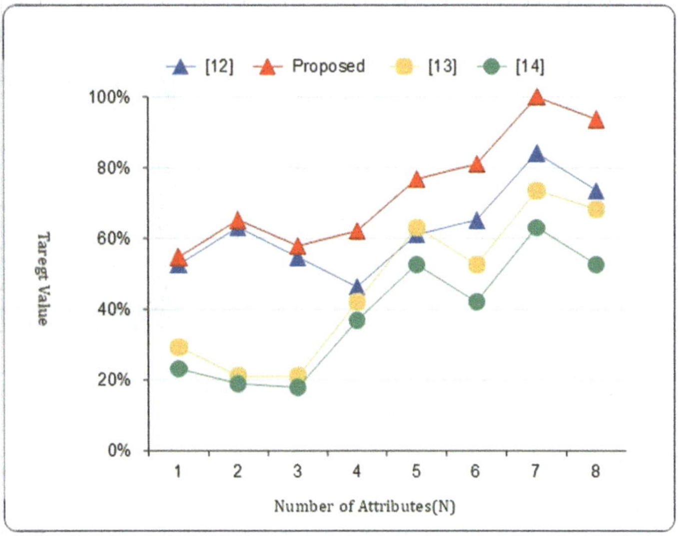

Fig. 3 Simulation results based on the number of attributes versus the target value of a number of packets dropped during transactions

virtual machine in the follower peer network has 30 vCPUs and 8 GB of RAM. The following procedure is carried out 30 times. "Second transaction" refers to the number of subsequent transactions handled every second (TPS). Responding times were between 445 milliseconds and 3867 milliseconds. The Hyperledger-based Intervened Time Fog Architecture is shown in Fig. 6. On the basis of how many qualities and the highest possible value were simulated, the results are shown in Fig. 13. A high figure indicates that a substantial number of packets were lost. In comparison to benchmark models, the number of packets lost in our suggested framework is lower. In order to compare the performance and efficiency of the proposed and benchmark models, we carried out a simulation based on a number of attributes and target values. From Fig. 3, it is very clear that our proposed framework with a minimum number of attributes performs efficiently, as can be seen from the graphs.

Figure 4 reveals the information related to the simulation results carried out on the basis of a number of attributes and a number of transactions. In Fig. 4, we have divided the number of attributes into clusters. Each cluster consists of 10 attributes. These attributes contain information related to name, type, address, geographic location, and insurance number or bank account. Based on the clustering techniques, we justify that our proposed framework performs better than the benchmark model.

In order to measure the performance validation of the proposed framework, we have carried out our experiments using the Hyperledger fabric blockchain simulation tool, as shown in Fig. 5. The parameters that were kept in this experiment are the number of attributes and the time complexity.

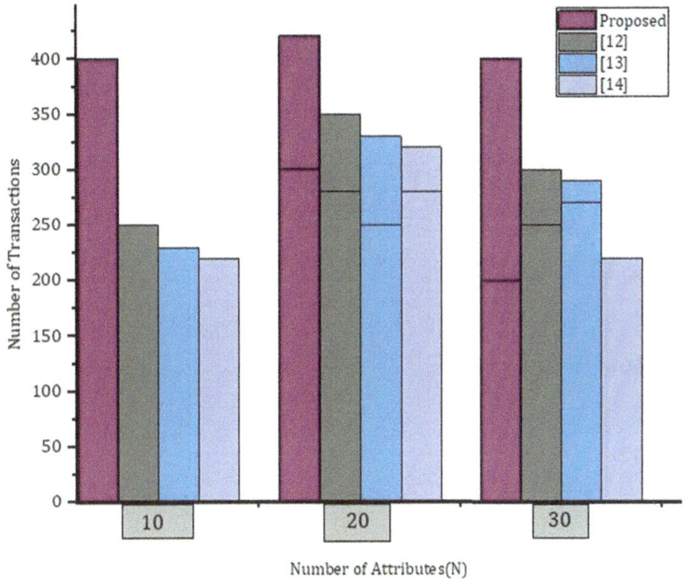

Fig. 4 Simulation results based on number of attributes versus number of transactions

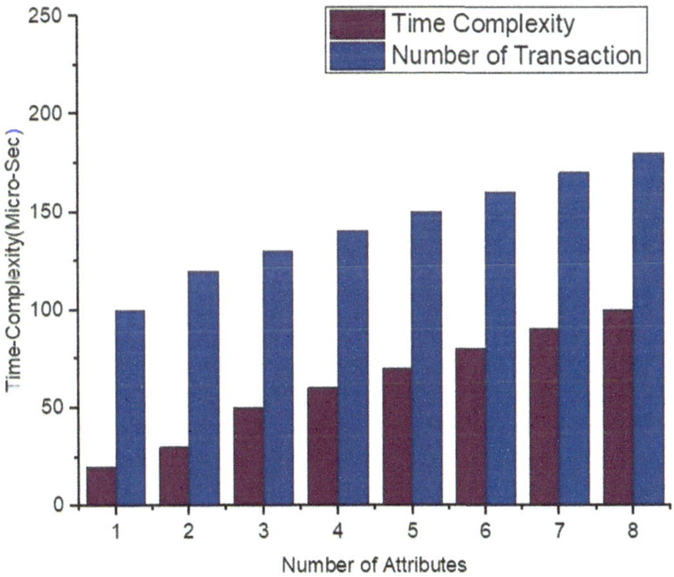

Fig. 5 Performance validations based on the experimental techniques using number of attributes versus time-complexity

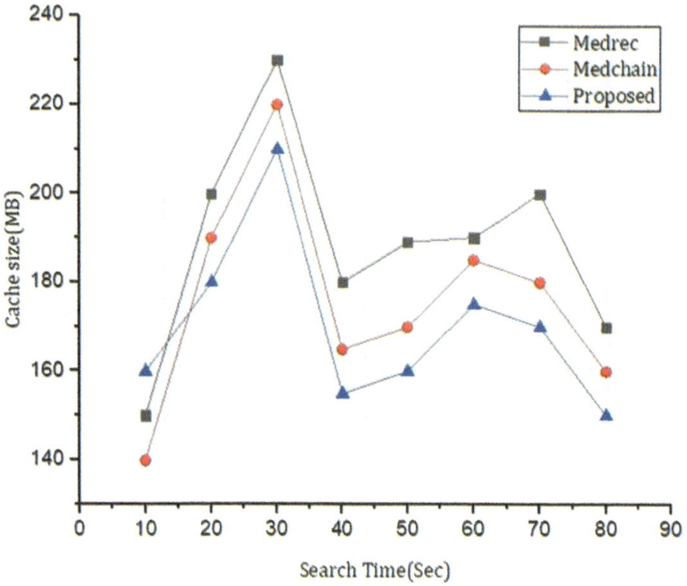

Fig. 6 Comparative results based on search time and cache hit size

$$\text{Throughput } (e) = \text{Number of Request } (N) \,/\text{Total Time } (T) \tag{1}$$

$$\text{Accuracy} = (\text{TP} + \text{TN}) \,/\, (\text{FP} + \text{FN} + \text{TP} + \text{TN}) \tag{2}$$

$$\text{Precision} = \text{TP} + \text{FP}/\text{TP} \tag{3}$$

$$\text{Recall} = \text{TP}/\text{FP} + \text{TN} \tag{4}$$

Figure 6 represents the comparative analysis of the proposed method versus the benchmark models based on the search time versus cache size.

Figure 7 represents the simulation results and comparative analysis based on the execution time versus cache hit rate (N). The more the cache hit rate achieved, the more will be the performance by the model.

In Fig. 8, the x-axis represents the number of rounds whereas the Y-axis represents the error rate. The simulation results in Fig. 8 describe the comparative analysis between our proposed model and the benchmark models. In this scenario, the performance parameters that we used to measure the performance are the number of rounds and the error rate. In order to measure the performance, the error rate should be less with a maximum number of rounds. From Fig. 8, it is very clear that the error rate is very less compared to the benchmark models. Therefore,

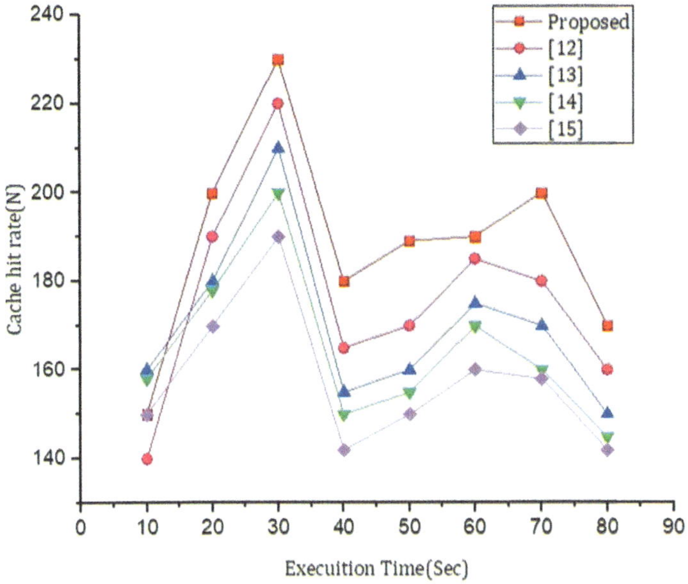

Fig. 7 Proposed secure searchable encryption system using DL based intrusion detection system

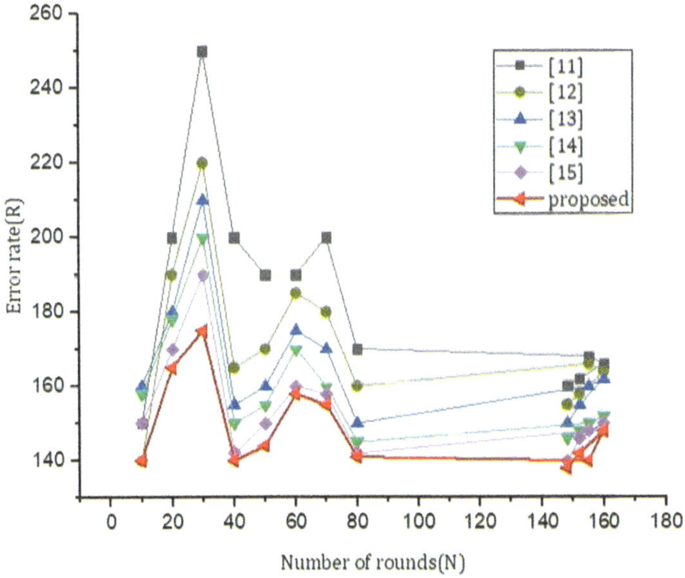

Fig. 8 Simulation results and comparative analysis based on the proposed framework versus benchmark models (error rate versus number of rounds)

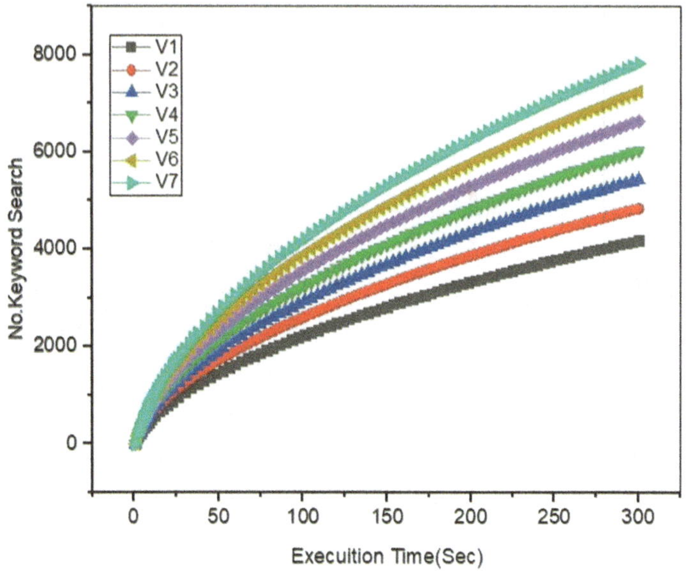

Fig. 9 Performance measurement based on simulation results using execution time versus number of keywords

Fig. 8 justifies that our proposed model receives 1.9× efficiency compared to the benchmark models. In Fig. 9, we have carried our experiment running on hyperledger fabric that reveals information based on DL and NN techniques by training the proposed models against intrusion detection. V1, V2..., and V7 represent the number of IoMT and Fog domains. V1 represent Fog1 and IoMT1. Similarly, V7 represents Fog domain 7 and IoMT7. The number of keywords searched during the experiments counts up to 8000 and the execution time up to 300 s. Figure 10 represent Computational cost analysis based on the sensors distance and the execution time for training the number of data-sets.

Figure 11 shows the performance evaluation based on NN and DL techniques. The performance is carried out through Fog with Blockchain (FB) and Fog without Blockchain (FwB) in terms of request arrival rate per second, number of requests per second, throughput generated by IoMT system according to our simulation, the current design can handle 200–2000 requests per second with a 200–1400 request arrival rate. The FogIoMT can handle 200–1400 arrival requests and finish 200–2000 requests per second. Figure 12 represents the simulations results through training the model using DL techniques. During the experiments, we have carried our 30% test data and 70% training data. From Fig. 12, it is very easy to justify as well as conclude that our proposed framework with the same training time has a less number of expected breaches as compared to the benchmark models. Figure 12 proves that our proposed framework provides more security as compared to the benchmark models (Fig. 13).

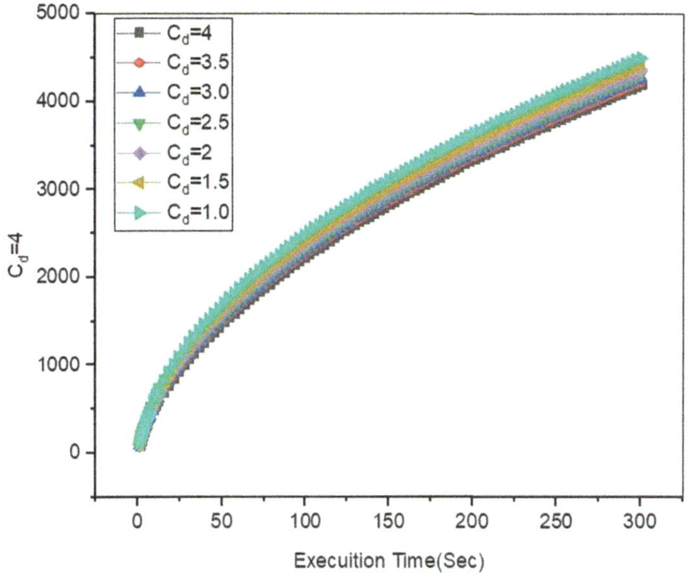

Fig. 10 Computational cost analysis based on the sensors distance and the execution time for training the number of datasets

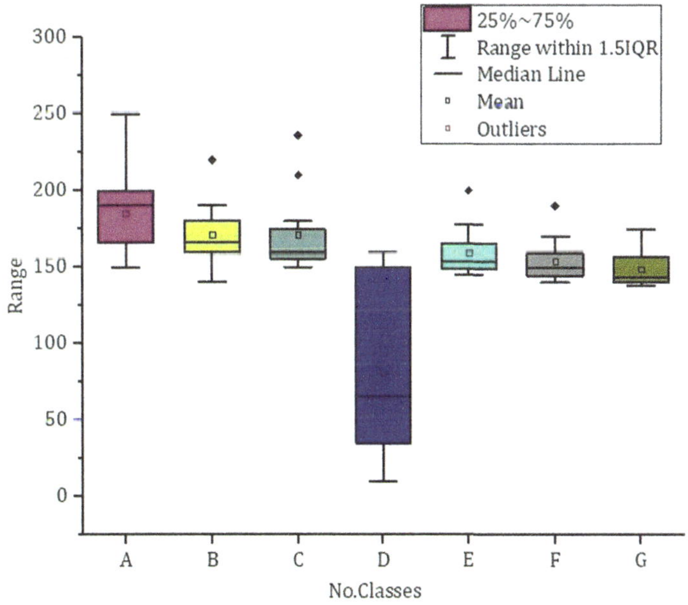

Fig. 11 Comparative analysis based on the number of rounds versus error rate

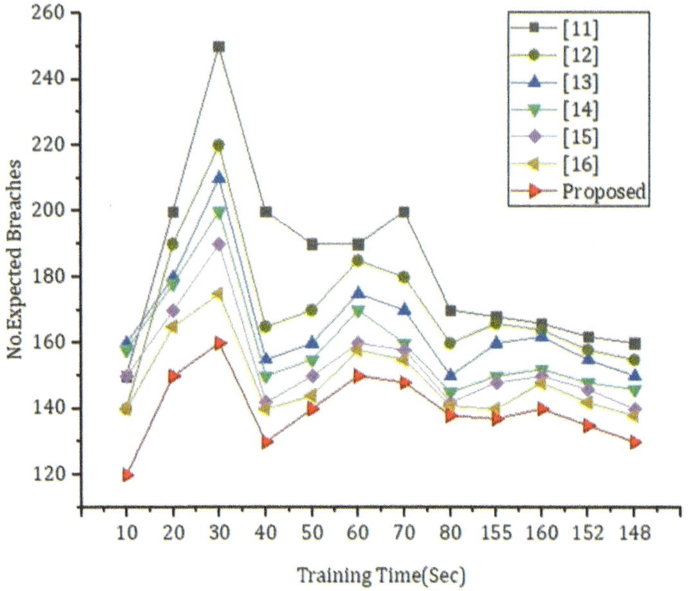

Fig. 12 Model training evaluation based on training time and number of expected breaches

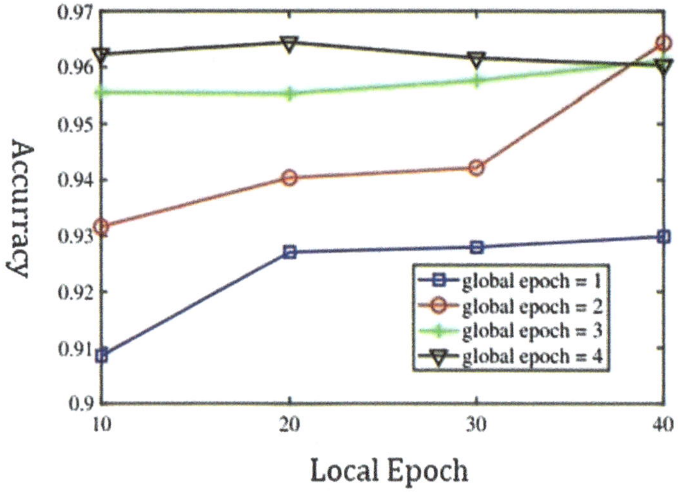

Fig. 13 Accuracy measurement and analysis using DL techniques using Fog with blockchain and Fog without blockchain

Figure 14 represents the simulations results based on the privacy parameters and test accuracy while we keep the variance in the global epoch. From the simulation results, it is very clear that the test accuracy for the global epoch 2 and 1 are obtained maximum by using the DL techniques. The model was trained by using real-time

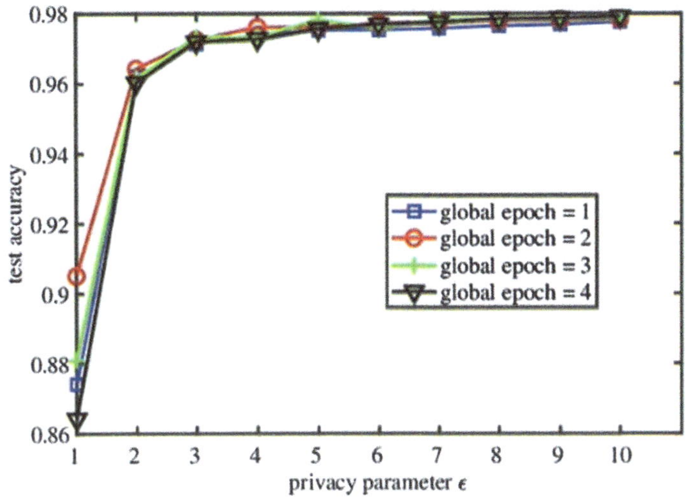

Fig. 14 Impact of DL on the test accuracy using our normalization technique under various global epochs

IoT-test net data-sets from the University of South-whale. From Fig. 14, it is very clear that the variant in the accuracy occurred when the privacy parameters reach 5 and 6.

4.1 Discussion and Analysis

A summary of our proposed work's experimental analysis is provided here. Aside from that, it provides a thorough explanation of the dataset used, experimental setup, and comparative analysis. It is possible to have several security concerns in an IoMT environment due to the above-mentioned dangers. PLCs and RTUs provide and receive IO signals from and to sensors and actuators in the field. Data of this nature should only be sent between specific sources and destinations under stringent control. Furthermore, this data is regarded as highly sensitive because it reveals the workings and logic of the control systems in question. Every device in the network will have access to the whole transaction history using the permissionless approach. This information will be made public if any of the devices are compromised. In the wrong hands, malevolent actors may be able to reverse engineer these machines and find ways to attack and interrupt their crucial operations. This means that the deployment of blockchain technologies such as consortium blockchain in highly sensitive and important situations such as the IIoT and the IoMT is difficult to justify. This is because, in exchange for their high integrity, blockchains expose users to greater risk. In Fig. 15, simulation results represent the comparative analysis of the proposed framework versus benchmark models. The comparisons are based on

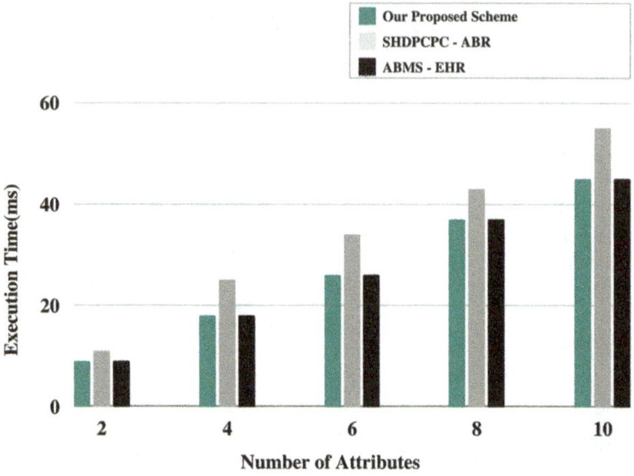

Fig. 15 Comparative analysis of the proposed approach versus benchmark models

number of transaction and d2d distance. Moreover, for the same number of distance between peer nodes, the number of transactions varies. Moreover, Fig. 15 provides the comparative analysis based on the network delay. It can be observed that the network delay for the proposed approach is less as compared to the benchmark approaches. The results presented in Fig. 15 are recorded to compare the proposed framework with the benchmark models. The parameters to evaluate the proposed framework are distances between two nodes and the number of transactions. Finally, Fig. 15 represents the comparative analysis of the proposed approach versus the benchmark models based on the number of attributes and execution time. The simulation results are based on the number of attributes (X-axis) and execution time (Y-axis). Moreover, it can be observed that using lightweight homomorphic encryption (HE), the proposed approach perform better than the benchmark models in terms of execution for the same number of attributes. In order to evaluate the attacks resistance of the proposed framework with the benchmark models, we carried out the comparison through Table 1.

5 Conclusion

One of the most hyped technologies in the last 5 years has been blockchain, thanks to the popularity it has earned due to its different crypto coins. Numerous blockchain-based use cases have been put into practice already, from financial transactions to identity management to supply chain management. There were, however, no use cases for critical infrastructure with sensitive systems and data as their assets in these examples. While blockchains such as Hyperledger fabric and Ethereum provide important anonymity, integrity, and audibility features for their users, there are important privacy and security risks that were discussed and presented in this chapter related to their use in critical environments such as IIoT environments and IOMT using Fog computing and cyberphysical systems. One of the primary architectural elements of a Blockchain is the distribution of the ledger. The privacy challenges highlighted in this research will be addressed in future developments of Hyperledger fabric blockchain.

References

1. A. A. Shah, G. Piro, L. A. Grieco, and G. Boggia, "A qualitative cross-comparison of emerging technologies for software-defined systems," in *2019 Sixth International Conference on Software Defined Systems (SDS)*, 2019, pp. 138–145.
2. A. Ali and M. Mehboob, "Comparative analysis of selected routing protocols for wlan based wireless sensor networks (wsns)," in *Proceedings of 2nd International Multi-Disciplinary Conference*, vol. 19, 2016, p. 20.
3. A. A. Shah, G. Piro, L. A. Grieco, and G. Boggia, "A review of forwarding strategies in transport software-defined networks," in *2020 22nd International Conference on Transparent Optical Networks (ICTON)*, 2020, pp. 1–4.
4. R. R. Bruce, J. P. Cunard, and M. D. Director, *From telecommunications to electronic services: A global spectrum of definitions, boundary lines, and structures*. Butterworth-Heinemann, 2014.
5. V. Gatteschi, F. Lamberti, C. Demartini, C. Pranteda, and V. Santamaría, "Blockchain and smart contracts for insurance: Is the technology mature enough?" *Future Internet*, vol. 10, no. 2, p. 20, 2018.
6. B. Jia, T. Zhou, W. Li, Z. Liu, and J. Zhang, "A blockchain-based location privacy protection incentive mechanism in crowd sensing networks," *Sensors*, vol. 18, no. 11, p. 3894, 2018.
7. K. Biswas and V. Muthukkumarasamy, "Securing smart cities using blockchain technology," in *2016 IEEE 18th international conference on high performance computing and communications; IEEE 14th international conference on smart city; IEEE 2nd international conference on data science and systems (HPCC/SmartCity/DSS)*. IEEE, 2016, pp. 1392–1393.
8. T. M. Fernández-Caramés, I. Froiz-Míguez, O. Blanco-Novoa, and P. Fraga Lamas, "Enabling the internet of mobile crowdsourcing health things: A mobile fog computing, blockchain and iot based continuous glucose monitoring system for diabetes mellitus research and care," *Sensors*, vol. 19, no. 15, p. 3319, 2019.
9. A. Ali, M. Naveed, M. Mehboob, H. Irshad, and P. Anwar, "An interference aware multichannel mac protocol for wasn," in *2017 International Conference on Innovations in Electrical Engineering and Computational Technologies (ICIEECT)*. IEEE, 2017, pp. 1–9.
10. A. Beebeejaun, "Vat on foreign digital services in Mauritius; a comparative study with South Africa," *International Journal of Law and Management*, 2020.
11. A. Aziz Shah, G. Piro, L. Alfredo Grieco, and G. Boggia, "A quantitative cross-comparison of container networking technologies for virtualized service infrastructures in local computing environments," *Transactions on Emerging Telecommunications Technologies*, vol. 32, no. 4, p. e4234, 2021.
12. A. Yazdinejad, R. M. Parizi, A. Dehghantanha, and K.-K. R. Choo, "Blockchain-enabled authentication handover with efficient privacy protection in sdn-based 5g networks," *IEEE Transactions on Network Science and Engineering*, 2019.
13. H. Kim, S.-H. Kim, J. Y. Hwang, and C. Seo, "Efficient privacy-preserving machine learning for blockchain network," *IEEE Access*, vol. 7, pp. 136 481–136 495, 2019.

14. A. Cirstea, F. M. Enescu, N. Bizon, C. Stirbu, and V. M. Ionescu, "Blockchain technology applied in health the study of blockchain application in the health system (ii)," in *2018 10th International Conference on Electronics, Computers and Artificial Intelligence (ECAI)*. IEEE, 2018, pp. 1–4.
15. A. Yazdinejad, G. Srivastava, R. M. Parizi, A. Dehghantanha, K.-K. R. Choo, and M. Aledhari, "Decentralized authentication of distributed patients in hospital networks using blockchain," *IEEE journal of biomedical and health informatics*, vol. 24, no. 8, pp. 2146–2156, 2020.
16. V. Patel, "A framework for secure and decentralized sharing of medical imaging data via blockchain consensus," *Health informatics journal*, vol. 25, no. 4, pp. 1398–1411, 2019.
17. Z. El-Rewini, K. Sadatsharan, D. F. Selvaraj, S. J. Plathottam, and P. Ranganathan, "Cybersecurity challenges in vehicular communications," *Vehicular Communications*, vol. 23, p. 100214, 2020.
18. A. Dorri, S. S. Kanhere, R. Jurdak, and P. Gauravaram, "Blockchain for iot security and privacy: The case study of a smart home," in *2017 IEEE international conference on pervasive computing and communications workshops (PerCom workshops)*. IEEE, 2017, pp. 618–623.
19. L. Hang and D.-H. Kim, "Design and implementation of an integrated iot blockchain platform for sensing data integrity," *Sensors*, vol. 19, no. 10, p. 2228, 2019.
20. B. Yu, S. K. Kermanshahi, A. Sakzad, and S. Nepal, "Chameleon hash time-lock contract for privacy preserving payment channel networks," in *International Conference on Provable Security*. Springer, 2019, pp. 303–318.
21. K. Hameed, A. Ali, M. H. Naqvi, M. Jabbar, M. Junaid, and A. Haider, "Resource management in operating systems-a survey of scheduling algorithms," in *Int. Conf. on Innovative Computing (ICIC)*, vol. 1. University Of Management and Technology, 2016.
22. A. D. Dwivedi, G. Srivastava, S. Dhar, and R. Singh, "A decentralized privacy-preserving healthcare blockchain for iot," *Sensors*, vol. 19, no. 2, p. 326, 2019.
23. E.-Y. Daraghmi, Y.-A. Daraghmi, and S.-M. Yuan, "Medchain: A design of blockchain-based system for medical records access and permissions management," *IEEE Access*, vol. 7, pp. 164 595–164 613, 2019.
24. Y. Jung, M. Peradilla, and R. Agulto, "Packet key-based end-to-end security management on a blockchain control plane," *Sensors*, vol. 19, no. 10, p. 2310, 2019.
25. C. Esposito, A. De Santis, G. Tortora, H. Chang, and K.-K. R. Choo, "Blockchain: A panacea for healthcare cloud-based data security and privacy?" *IEEE Cloud Computing*, vol. 5, no. 1, pp. 31–37, 2018.
26. C. W. Choo, *Information management for the intelligent organization: the art of scanning the environment*. Information Today, Inc., 2002.
27. S. K. Kermanshahi, J. K. Liu, R. Steinfeld, S. Nepal, S. Lai, R. Loh, and C. Zuo, "Multi-client cloud-based symmetric searchable encryption," *IEEE Transactions on Dependable and Secure Computing*, 2019.
28. S. K. Kermanshahi, J. K. Liu, and R. Steinfeld, "Multi-user cloud-based secure keyword search," in *Australasian Conference on Information Security and Privacy*. Springer, 2017, pp. 227–247.

Detecting Collusive Agents by Trust Measures in Social IoT Environments: A Novel Reputation Model

Mariantonia Cotronei, Sofia Giuffrè, Attilio Marcianò, Domenico Rosaci, and Giuseppe M. L. Sarnè

1 Introduction

The world around us is quickly changing, thanks also to the Internet of Things (IoT) that, as predicted by Kevin Ashton in 1994 [6], is growing faster and faster at all levels of our daily life.

Indeed, the idea behind IoT is to make smart objects equipped with a network interface and of both sensing and reasoning capabilities in order to provide them with communication and cooperation skills [18, 33]. Through network infrastructures, IoT objects are identifiable by means of an IP address, traceable in time and space, and can potentially interact with millions of other IoT objects and people. At the same time, data and services can be generated, aggregated, and customized according to the needs of other smart objects and/or human users [5] which, in turn, can exploit the opportunities to dialogue with sensors and actuators. As a result, IoT objects generate multidimensional and context-sensitive smart network infrastructures potentially rich of social interactions [37].

From social and technological viewpoints, applying a social structure to real IoT objects requires the development of suitable paradigms to manage the interactions that an IoT object can carry out with the environment in which it lives [7]. In this context, the social IoT (SIOT) studies social interactions happening in IoT environments including service-oriented interactions (e.g., marketing, health care,

M. Cotronei · S. Giuffrè · A. Marcianò (✉) · D. Rosaci
DIIES - Università Mediterranea of Reggio Calabria, Reggio Calabria, Italy
e-mail: mariantonia.cotronei@unirc.it; sofia.giuffre@unirc.it; attilio.marciano@unirc.it; domenico.rosaci@unirc.it

G. M. L. Sarnè
Department of Psychology, University of Milan Bicocca, Milan, Italy
e-mail: giuseppe.sarne@unimib.it

L. Fotia et al. (eds.), *Security, Trust and Privacy Models, and Architectures in IoT Environments*, Internet of Things, https://doi.org/10.1007/978-3-031-21940-5_3

home automation, security, information, transport, etc.) that can be provided upon a payment or for free [35].

An effective intuition to boost these social interactions consists of associating IoT smart objects with software agents which work on their behalf. The social attitude of software agents to interact and collaborate enables us to realize complex and sophisticated forms of collaboration among IoT objects [7]. Moreover, the convergence between both the IoT and the agent technologies allows to overcome also strong forms of heterogeneity among IoT objects, enabling scalable and pervasive cooperation and interactions among them, as well as service discovery and/or composition [49]. Note that in the following, the terms smart object, IoT object, IoT device, and agent will be used in an interchangeably way when not specified otherwise.

Furthermore, thereafter we will assume IoT interactions qualitatively analogous to those occurring in human societies considering actors (i.e., IoT objects) with human-like characters (e.g., benevolent, selfish, honest, malicious, etc.) and playing analogous roles (e.g., provider, consumer, prosumer). Therefore, from a SIOT perspective, the presence of malicious actors in IoT environments is real, and it should be considered. These malicious agents act to obtain undue benefits, by performing cheating activities by means of a wide range of different (and not infrequently concurrent) malicious behaviors, from simple ones (e.g., selfishness, misjudgment, etc.) to more complex ones (e.g., collusive, alternate, etc.). In addition, IoT devices are exposed to greater risks of malicious attacks in presence of large IoT environments, nomadic IoT objects or other situations that might increase the probability of interacting with partners whose trustworthiness is still unknown [17]. Obviously, minimizing such risks is of primary importance to enhance IoT social aspects and user acceptance, as well as for promoting the consumption of IoT services and applications.

An effective solution to minimize risks is to provide each player with appropriate trust measures about potential partners. This can improve the probability of making the best possible partner choice or, conversely, deciding not to interact with anyone. However, from a broader perspective, trust is complex to define and measure because it is influenced by measurable and not measurable properties and refers, for instance, to safety, goodness, availability, capability, and other attributes. Therefore, due to this multifaceted nature, several meanings can be associated with the term "trust."

Two trust aspects are widely considered by researchers with respect to the *reliability trust* and the *decision trust*. To this purpose, a large number of definitions have been provided about them in the literature. For instance, for the former we can cite the well-known definition provided by Gambetta [20] "Trust is the subjective probability by which an individual, *A*, expects that another individual, *B*, performs a given action on which its welfare depends"; while for the second one, McKnight and Chervany [41] assert that "Trust is the extent to which one party is willing to depend on something or somebody in a given situation with a feeling of relative security, even though negative consequences are possible." The trust definitions provided above are aimed to address subjectivity and situational risks, two aspects intrinsic in trust and playing a main role in the IoT.

Supporting the agents in their activities, overcoming the uncertainty perceptions, and mitigating the risk of deception, a great number of *Trust and Reputation Systems* (TRSs) have been proposed in the literature, categorized on the basis of the particular used approach (hierarchical, reputation based, derived from social networking, fuzzy techniques, past behavior history, etc.) [4, 16, 56]. TRSs can be assumed as decision support tools, which are widely adopted in electronic communities to assess the trustworthiness of their members, usually by means of a synthetic numerical score. More specifically, when a community member desires to compute the trust score of another member (i.e., the trustee), both direct and/or indirect information could be used. In the first case, the information derives by a direct knowledge of the trustee; the other one consists of ratings and/or opinions provided by the other members (i.e., the trustors) of the own community about the trustee. Then the computed trust score can be conveniently used to detect dishonest actors and/or select the best counterpart to interact with.

More specifically, the contribution we present in this paper consists of:

- Proposing a method to preliminarily identify the best candidates as malicious (colluding) in order to use such agents as pre-untrusted entities.
- presenting a novel reputation model to detect colluding malicious agents; our approach, specifically designed for IoT agent-based social communities, does not penalize reputation scores of honest agents and keeps them unchanged, unlike other well-known TRSs [31].

The rest of the paper is organized as follows. Section 2 presents the related work. In Sect. 3 the reference IoT scenario is introduced. The proposed reputation model is presented and discussed in Sect. 4, while two cases study are illustrated in Sect. 5. Finally, in Sect. 6 some conclusions are drawn.

2 Related Work

In social activities (communications, multi-agent systems, peer-to-peer activities, and many other areas [3, 13, 18, 26, 47]), the capability to identify the trustworthiness of an actor, be it a human, a software or a device, is of primary relevance for detecting malicious actors and to marginalize or expel them from the community. To assess the trustworthiness of an entity, it is possible to leverage on opinions raised from the perspective of own past direct experiences with that entity (i.e., obtaining a measure of "reliability") or/and aggregating feedback provided by other members of the community (i.e., obtaining a measure of "reputation"), otherwise generally referred as indirect experiences [11].

As previously specified, *Trust and Reputation Systems* (TRSs) are tools designed to support a "shield" against malicious actors that might perform various misleading behaviors [15, 39, 43], even simultaneously, to obtain undue benefits. In more detail,

Trust Systems (TSs) consider both direct and indirect information sources, combined by adopting different rules, e.g., based on the level of direct knowledge that might also vary over time [46]. Differently, *Reputation Systems* (RSs) rely only on indirect information sources [23]. However, as a result, TRSs compute a score (generally unique) to appropriately and synthetically describe the trustworthiness of an actor.

A large number of TRSs, for general purpose or specialized for well-defined contexts, have been designed with the aim to be resilient to malicious attacks [9, 24]. We can argue that the greater the degree of TRSs robustness to malicious attacks, the more reliable its trustworthiness measures will be [27]. In addition, we might also consider the percentage of recognized cheaters as a possible quantitative measure of the TRSs robustness or, in other words, of their effectiveness.

Preliminary, note that the TRSs encompass systems that, while representing comparable concepts, inherently differ for the nature of the information sources involved, for the centralized or decentralized architecture adopted or for the adoption of a global or local approach (e.g., based on individual ego-networks), etc. [12, 29, 50]. In the following, the focus will be given only on the capability of TRSs to identify malicious actors regardless the characteristics above highlighted.

A significant number of researches have compared the robustness of TRSs with respect to a more or less wide range of malicious behaviors, although it may be difficult to compare systems designed to work in different contexts. An earlier analysis of TRSs is provided in [24], where the defense mechanisms of some TRSs with respect to the most common malicious attacks are compared. Other similar studies can be found in [28, 52]. In particular, a common characteristic of these studies is the absence of a well-defined quantitative approach to assess the strengths and weaknesses of TRSs.

After these initial attempts, special simulated scenarios, populated by both malicious and honest agents, were implemented to test TRSs. These simulations were aimed at realizing some forms of competition to compare a TRS proposal with other competitors in order to measure their performance and/or some property, as in [34, 54]. Then, some authors proposed to implement specific test beds to systematically compare different TRSs approaches and test them under the same conditions. Among them, the well-known ART (Agent Reputation and Trust test bed) [19] has found wide acceptance in the literature, and in the past it has also been used for a TRSs competition that was held annually. ART simulates a picture auction, and the aim of the ART participants was to gain a measurable economic profit. Other examples of test beds can be found in [2, 32, 40].

The main problem with the simulations is that the worst-case scenario for a TRS may neither be realized nor recognized autonomously. This is due to the fact that TRSs are generally designed for specific scenarios and relocating them to other contexts can lead to incorrect assessments. As an alternative to test beds, a different and more complex methodology to verify the robustness of TRSs requires the use of mathematical/analytical approaches [8, 21]. Their advantage is in the possibility of a more comprehensive verification of the strengths and limitations of a TRS, but, conversely, they are not directly transferable from a TRS to another.

Examining some TRSs in more detail, one of the most popular and studied is that used by eBay [22]. More specifically, it is a quite simple RS, and while its simplicity has met with user acceptance, on the other hand, this does not provide it with a particular robustness to malicious attacks [10, 22, 45]. Therefore, it has been updated several times but without changing its basic characteristics. The eBay RS simply takes into account the number of positive, neutral, and negative feedback received by the counterparts in few time windows. All newcomers are assigned a null reputation score. A peculiarity of this RS is that it is left to the users to evaluate the trustee's reputation on the basis of their risk attitude.

Based on a distributed architecture and a peer-to-peer overlay network, PeerTrust [55] is a TRS resilient to different types of malicious attacks that on the basis of computed reputation scores, allows to identify those peers considered as more suitable to interact. Such a robustness derives by the combination of different information such as the specific context, the direct feedback, the credibility of the indirect sources, and the number of transactions performed by each peer and transaction information. Another decentralized TRS that, in large competitive federations of utility computing infrastructures, operates using an overlay network linking all federated nodes is Hypertrust [42]. In Hypertrust, the nodes are interconnected via overlay links and form clusters; then a distributed algorithm operates to discover and allocate resources associated with trusted nodes. To limit the resource search to an eligible region, Hypertrust assesses information related to both trustworthiness and reputation (obtained in the form of recommendations) and allows each node to implement an efficient search for resources potentially of its interest.

Always in a peer-to-peer context, EigenTrust [31] is one of the most well-known and studied reputation algorithm. EigenTrust, which can be seen as a variation of the famous PageRank algorithm, aggregates into a trust matrix all the (normalized) ratings, called local trusts, that each peer has about the trustworthiness of every other peer in the community. Then the local trust scores are weighted on the basis of the trustworthiness of each trustor peer. These normalized scores are used by the EigenTrust algorithm to compute the global reputation of each peer by making the assumption of reputation transitivity.

A significant number of TRSs are based on assumptions unfitting with open environments. For example, the FIRE TRS [25], operating in a multi-agent context, considers the nature of all agents to be always honest and benevolent (e.g., on-demand agents exchange correct information). Fire computes agents' trustworthiness based on multiple information sources, i.e., *interaction trust*, *role-based trust*, *witness reputation*, *feedback source*, and *certified reputation*. These information sources are respectively referred to (*i*) the direct experience of the agent, (*ii*) the nature of agents' relationships, (*iii*) the source releasing the feedback about the behaviors of other agents, and (*iv*) the reliability of the agent suggested by the rated agent as third party. Unfortunately, the effects of the initial assumptions make this TRS susceptible to malicious attacks such as collusion, alternation, complaining, and others [48]. Moreover, FIRE needs to set a large number of parameters which represents another critical issue of this TRS.

In the IoT world, the problem of recognizing malicious agents can be more complex than compared to other contexts. In fact, additional constraints in terms of computational, memory, and power resources of the IoT devices or the possibility of their frequent migrations between different environments must be taken into account in designing a TRS for the IoT. However, an increasing number of TRSs for modeling trust in distributed, semi-distributed, centralized, and/or blockchain-based IoT environment are proposed in the literature [1, 4, 14, 16].

In open and dynamic IoT environments, security and privacy are significant issues due to the increased opportunity for anomalous behaviors. Approaches purely based on authentication methods may not be feasible and/or inadequate to build trust. In [13], a distributed reputation model and a framework to manage the reputation of IoT devices are proposed. This model considers some countermeasures to detect malicious or cheating actors. A set of experiments, simulating vehicular mobility, on a basic urban transportation network, has been carried out by testing the effectiveness of the reputation model in quickly identifying malicious.

RESIOT [18] is a framework to form communities of reliable IoT devices based on the reputation of agents associated with IoT devices, to take advantage from the social attitude of agents. RESIOT leverages the well-known concept of social resilience applied to IoT systems. In this proposal, a novel reputation model is presented, and a series of experiments were performed in a simulated environment, where malicious agents conduct attacks following different, concomitant cheating strategies. The authors have shown that the proposed approach performs better than the other RSs exploited as competitors.

In [36], an approach, called Perceptron Detection, is described. It adopts a perceptron neural network and a K-means clustering method to support the computation of trust values and detect malicious nodes that can perpetrate three popular attack modes, i.e., tamper, drop, and replay attacks in both single and multiple modalities. Finally, to enhance performance, a suitable learning process has been designed by the authors. For a SIOT scenario, [38] proposes a TS adopting machine learning techniques. This TS has shown to be resilient to a significant number of attacks and detect almost all malicious nodes in the network at the number of required transactions increases.

Finally, to overcome the aforementioned IoT constraints, a number of proposals, mainly agent-based, adopt edge and/or cloud computing to move the most part of computational, storage, and communication loads into such environments, freeing IoT devices from such tasks. To guarantee fairness and traceability, a distributed blockchain-based trust modeling process is proposed for malicious node detection in wireless sensor networks [51]. Malicious node detection is accomplished by locating the sensor position and the use of smart contracts. Wang et al. [53] introduce a new trust evaluation model that employs trust transitivity on a chain assisted by mobile edge nodes to ensure the trustworthiness of IoT nodes and prevent malicious attacks. Mobile edge nodes offer an attractive solution by enabling relatively high compute and storage capacities compared to IoT devices. In the proposal, different computational approaches are provided for different chains of trust in order to measure their trust degrees. Second, through a modified Dijkstra algorithm, the trust information of sensor nodes from mobile edge nodes is collected. Experimental results show that this approach is more accurate and improve security in IoT.

To summarize, some TRSs have been presented in this section, which considered different and effective approaches to identify malicious actors both IoT and non-IoT domains. However, to the best of our knowledge any TRS, including those presented in this section, is explicitly designed to not alter the trustworthiness measures of honest in looking for malicious. Differently, the TRS presented in Sect. 4 accurately takes care also for this ethical and practical aspect.

3 The IoT Agent-Based Reference Scenario

In this section, we describe the IoT reference scenario which we will refer to in the following. This scenario involves a potentially large number of IoT devices, which are assisted by personal software agents. In such a community, the agents can mutually interact among them, on behalf of their associated devices.

In particular, within this agent community, we assume that the interactions carried out by agents satisfy the following desirable properties [44]:

- All the agents belonging to our community are long living entities so that their past behaviors provide us with information about their expected, future behaviors.
- When an agent decides to perform a new interaction with another agent, its choice is driven only by the past behaviors of its potential counterpart.
- The reputation of an agent must be spread into the community so that every other agent has the necessary information to make an informed choice about potential partners to interact with.

4 The Proposed Reputation Model

In the scenario above described, let $\mathcal{F}$ be the framework of the personal agents, each one associated with an IoT device, and let n be the number of such agents (i.e., IoT devices). Moreover, we assume that each agent $a \in \mathcal{F}$ is unambiguously identified as a_i, with $i \in [1, n]$.

Let t_{ij} be a real number, $0 \le t_{ij} \le 1$ representing the *trust* perceived by a_j about a_i. We assume $t_{ii} = 0$, with $i = 1, \ldots, n$; in other words, the trust of an agent about itself is not considered in our model.

The *reputation* of each agent in $\mathcal{F}$ is computed as a weighted sum. More in detail, let a_i be an agent of $\mathcal{F}$ (i.e., trustee); its reputation r_i in $\mathcal{F}$ is computed as:

$$r_i = \frac{\sum_{j=1}^{n} t_{ij} r_j}{\sum_{j=1}^{n} r_j}, \qquad i = 1, \ldots, n \tag{1}$$

So the trust of the agent a_i is given by the ratio between the sum of the trust values t_{ij} that the agents a_j have about a_i (i.e., trustor), with $j = 1, \ldots, n$, weighted by their own reputation scores, and the sum of the reputation of all such trustor agents

Now, we denote with **T** the trust matrix:

$$\mathbf{T} = \begin{pmatrix} t_{11} & t_{12} & \cdots & t_{1n} \\ t_{21} & t_{22} & \cdots & t_{2n} \\ \vdots & \vdots & \ddots & \vdots \\ t_{n1} & t_{n2} & \cdots & t_{nn} \end{pmatrix}.$$

T can be assumed as the transpose of the weighted adjacency matrix **A** corresponding to a directed graph where nodes correspond to the agents. Moreover, each link (i, j) is associated with a nonnegative value representing the trust value that the agent a_i perceives about the agent a_j.

Consequently, the Eqs. (1) can then be rewritten as:

$$\mathbf{Tr} = \mathbf{r}, \qquad \|\mathbf{r}\|_1 = 1, \tag{2}$$

where the i-th element of the vector $\mathbf{r} = (r_1, \ldots, r_n)^T$ is the reputation of the agent a_i, while the 1-*norm* is the sum of the absolute values of the elements of the vector $\mathbf{r}$. Note that $\|\mathbf{r}\|_1 = 1$ is required in order to guarantee the uniqueness of the solution to (2).

Moreover, we also impose that the sum of the trust value t_{ij} that each agent a_j assigns to the other agents of $\mathcal{F}$ is 1, i.e.:

$$\sum_{i=1}^{n} t_{ij} = 1. \tag{3}$$

As a result, the matrix **T** is column-stochastic.

To solve the eigensystem problem (2), we observe that it can be equivalently reformulated as the computation of the stationary distribution for a Markov chain represented by the transition matrix **T**.

Consequently, from the Perron Frobenius theorem, $\lambda = 1$ is the largest eigenvalue of **T**. Furthermore, as it is well-known, under the assumption $t_{ij} > 0$, there exists a unique vector $\mathbf{r} \in \mathbb{R}$, $\|\mathbf{r}\|_1 = 1$, such that $\mathbf{Tr} = \mathbf{r}$. The PageRank model is a modified version of the eigensystem (2). More formally:

$$\left(\delta\,\mathbf{T} + (1-\delta)\,\mathbf{p}\,\mathbf{q}^T\right)\mathbf{r} = \mathbf{r} \tag{4}$$

where the parameter $\delta \in \mathbb{R}$ ranges in $[0, 1]$ and $\mathbf{q} = (1, 1, \ldots, 1)^T$ and $\mathbf{p}$ (generally named *teleportation vector*) is a nonnegative vector with unitary 1-norm, i.e., $\mathbf{q}^T\mathbf{p} = 1$. If $\delta \neq 0, 1$, the solution to (4) exists and it is unique.

In the original PageRank algorithm, all the elements of the vector $\mathbf{p}$ are set to $1/n$. In [30], the authors propose to consider some *pre-trusted agents*, i.e., some a priori chosen agents whose opinions are considered highly reliable. Let $\mathcal{K}$ be the set of such agents so that $p_i = 1/|\mathcal{K}|$ if the agent belongs to $\mathcal{K}$ and 0 otherwise.

To detect collusive malicious agents, we propose to compute the vector $\mathbf{r}$ more realistically from the information directly available in the matrix $\mathbf{T}$.

For this purpose, any pair of agents (a_i, a_j) is considered malicious when all of the following conditions are simultaneously satisfied:

- The trust values t_{ij} and t_{ji} are large.
- The trust values t_{ij} and t_{ji} are similar.
- The sum of the remaining trust scores from rows i and j, respectively associated with agents a_i and a_j, is small.

In other words, we pay attention to all those cases where the trust values t_{ij} and t_{ji} of agents a_i and a_j are large, while the majority of the other agents of $\mathcal{F}$ assign low trust values to them.

Formally, consider a vector $\mathbf{s}$ representing the out-degree of the nodes of the associated graph or, similarly, the vector of the sums of the rows of $\mathbf{T}$, i.e., $\mathbf{s} = \mathbf{T}\,\mathbf{q}$. Furthermore, we fix three threshold values $\alpha, \beta, \gamma \in \mathbb{R}$ over $[0, 1]$.

Let $\tilde{\mathbf{E}}$ be an auxiliary matrix built from the matrix $\mathbf{T}$ by identifying the quasi-symmetric high-valued elements as follows:

$$\tilde{e}_{ij} = \begin{cases} t_{ij}, & \left|t_{ij} - t_{ji}\right| \leq \alpha \text{ and } t_{ij} \geq \beta \\ 0 & \text{otherwise} \end{cases} \tag{5}$$

The matrix $\tilde{\mathbf{E}}$ corresponds to a weighted adjacency matrix of the (undirected) subgraph whose arcs connect potential colluding agents.

Then the vector $\tilde{\mathbf{s}} = \mathbf{s} - \tilde{\mathbf{E}}\,\mathbf{q}$ is built. Based on the vector $\tilde{\mathbf{s}}$, we classify as malicious an agent belonging to $\mathcal{F}$ if:

$$\tilde{s}_i \leq \gamma$$

The knowledge of supposed malicious agents can be exploited in a variety of ways to provide appropriate reputation vectors.

For instance, let $\mathcal{S}$ be the index set of the malicious agents previously identified, and, considering the graph associated with $\mathcal{F}$, let $\mathbf{Z}$ the definitive weighted adjacency matrix associated with the subgraph consisting of all the both colluded agents and connecting edges.

For the first one case, consider (4) and a vector **v**, whose elements v_i are set to 0 if $i \in \mathcal{S}$ or $1/(n - |\mathcal{S}|)$ otherwise. By adopting this setting, all the honest agents will receive the same trust score regardless from their starting trust score. This is exactly the same algorithm as EigenTrust; however, we exploit here the additional information about the pre-trusted users.

The second one approach consists of building a new matrix $\tilde{\mathbf{T}}$ as follows:

- Fix a value $\varphi > 0$.
- For $a_i, a_j \in \mathcal{S}$; then let $\tilde{t}_{ij} = \varphi$, otherwise $\tilde{t}_{ij} = t_{ij}$.
- Make the $\tilde{\mathbf{T}}$ columns stochastic by normalizing each of them to 1.

5 Two Cases Study

To illustrate the effectiveness of the reputation system that we conceived, in the following, we present two simple examples.

5.1 Case Study 1

Consider an IoT community formed by six IoT devices (i.e., agents) that is represented in Fig. 1.

Moreover, we assume that suspected malicious agents are the agents identified as a_3 and a_4.

The corresponding trust matrix is:

$$\mathbf{T} = \begin{pmatrix} 0 & 0.75 & 0.001 & 0.003 & 0.43 & 0.45 \\ 0.76 & 0 & 0.002 & 0.001 & 0.46 & 0.45 \\ 0.003 & 0.001 & 0 & 0.87 & 0.001 & 0.002 \\ 0.002 & 0.003 & 0.88 & 0 & 0.002 & 0.003 \\ 0.135 & 0.126 & 0.015 & 0.08 & 0 & 0.095 \\ 0.1 & 0.12 & 0.1 & 0.046 & 0.107 & 0 \end{pmatrix}.$$

Analyzing the trust matrix T and applying our threshold strategy (5) in correspondence of $\alpha = 0.03$ and $\beta = 0.45$, the weighted adjacency matrix that identify the potential colluded agents is:

$$\tilde{\mathbf{E}} = \begin{pmatrix} 0 & 0.75 & 0 & 0 & 0 & 0 \\ 0.76 & 0 & 0 & 0 & 0 & 0 \\ 0 & 0 & 0 & 0.87 & 0 & 0 \\ 0 & 0 & 0.88 & 0 & 0 & 0 \\ 0 & 0 & 0 & 0 & 0 & 0 \\ 0 & 0 & 0 & 0 & 0 & 0 \end{pmatrix}$$

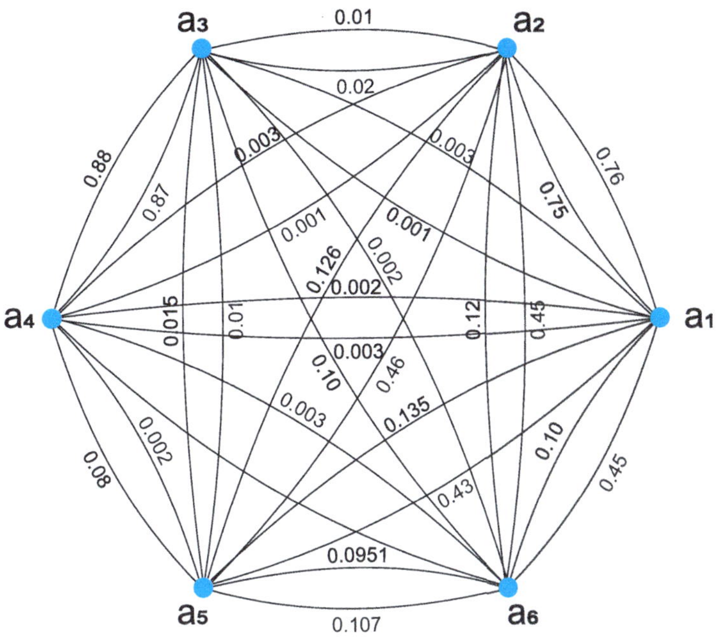

Fig. 1 Case study A, example of IoT community formed by 6 IoT devices (i.e., agents)

The choice of $\gamma = 0.45$ confirms that the only malicious colluding agents are the devices 3 and 4.

Once the malicious agents have been identified, we can illustrate the two approaches described in Sect. 4.

Let $\mathcal{S} = \{3, 4\}$; we obtain the vector $\mathbf{v} = (0.25, 0.25, 0, 0, 0.25, 0.25)^T$.

Then we set the parameter δ to 0.2.

Following the first approach A, the updated trust matrix is computed as:

$$\tilde{\mathbf{T}}_{\mathbf{A}} = \begin{pmatrix} 0.2000 & 0.3500 & 0.2002 & 0.2006 & 0.2860 & 0.2900 \\ 0.3520 & 0.2000 & 0.2004 & 0.2002 & 0.2920 & 0.2900 \\ 0.0006 & 0.0002 & 0 & 0.1740 & 0.0002 & 0.0004 \\ 0.0004 & 0.0006 & 0.1760 & 0 & 0.0004 & 0.0006 \\ 0.2270 & 0.2252 & 0.2030 & 0.2160 & 0.2000 & 0.2190 \\ 0.2200 & 0.2240 & 0.2200 & 0.2092 & 0.2214 & 0.2000 \end{pmatrix}$$

Finally, by adopting our new method B and by choosing $\varphi = 0.00033$, the trust matrix will be:

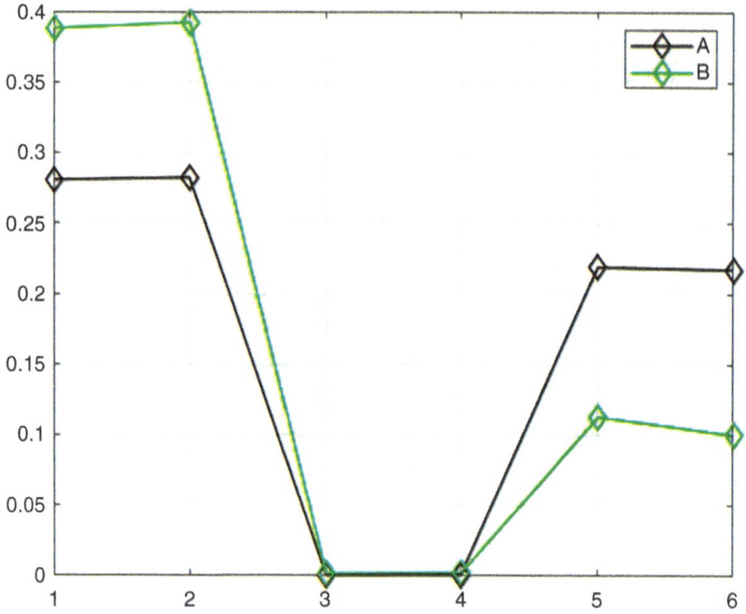

Fig. 2 The reputation vectors obtained with the approaches A and B

$$\tilde{\mathbf{T}}_{\mathbf{B}} = \begin{pmatrix} 0 & 0.7500 & 0.0085 & 0.0230 & 0.4300 & 0.4500 \\ 0.7600 & 0 & 0.0169 & 0.0077 & 0.4600 & 0.4500 \\ 0.0030 & 0.0010 & 0 & 0.0025 & 0.0010 & 0.0020 \\ 0.0020 & 0.0030 & 0.0028 & 0 & 0.0020 & 0.0030 \\ 0.1350 & 0.1260 & 0.1268 & 0.6138 & 0 & 0.0950 \\ 0.1000 & 0.1200 & 0.8451 & 0.3530 & 0.1070 & 0 \end{pmatrix}$$

In Fig. 2 the reputation vectors corresponding to cases A and B are plotted.

It is clear that both approaches penalize the malicious agents. Moreover, the approach A rewards all honest agents $(1, 2, 5, 6)$ almost equally, whereas our approach B mainly promotes the good honest agents $(1, 2)$.

5.2 *Case Study 2*

As a further case study, we show how the choice of γ is fundamental to identify only the malicious colluded agents and to assign the correct reputations.

In this case, we consider an IoT community formed by five IoT devices (i.e., agents) represented in Fig. 3, where:

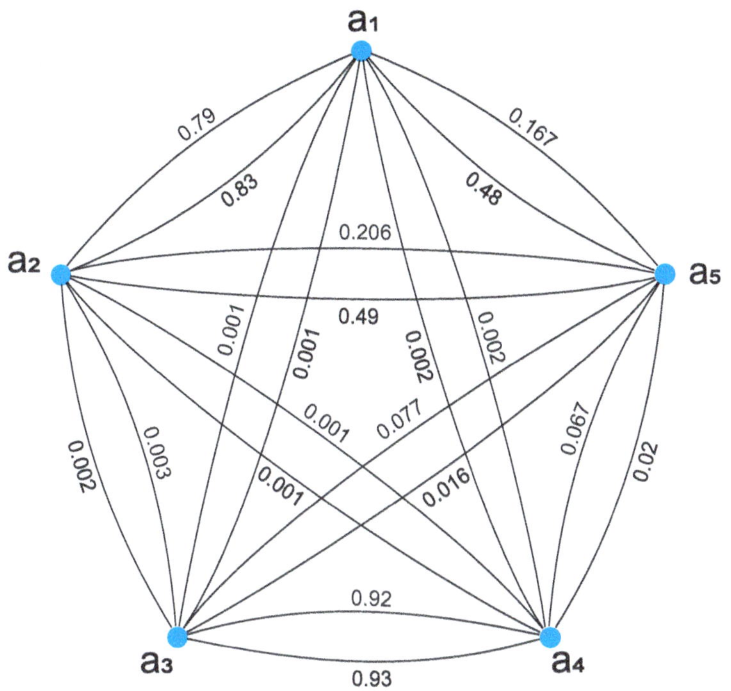

Fig. 3 Case study B, example of IoT community formed by 5 IoT devices (i.e., agents)

- The agents a_1 and a_2 are good honest agents that receive high values from all honest agents.
- The agents a_3 and a_4 are malicious colluded agents that exchange high values with each other and they attribute low values to the others:
- The agent a_5 is simply an honest agent.

The corresponding trust matrix is:

$$\mathbf{T} = \begin{pmatrix} 0 & 0.79 & 0.001 & 0.002 & 0.48 \\ 0.83 & 0 & 0.002 & 0.001 & 0.49 \\ 0.001 & 0.003 & 0 & 0.93 & 0.01 \\ 0.002 & 0.001 & 0.92 & 0 & 0.02 \\ 0.167 & 0.206 & 0.077 & 0.067 & 0 \end{pmatrix}$$

Applying our threshold strategy (5) in correspondence of $\alpha = 0.04$ and $\beta = 0.49$ the weighted adjacency matrix that identify the potential colluded agents is:

$$\tilde{\mathbf{E}} = \begin{pmatrix} 0 & 0.79 & 0 & 0 & 0 \\ 0.83 & 0 & 0 & 0 & 0 \\ 0 & 0 & 0 & 0.92 & 0 \\ 0 & 0 & 0.93 & 0 & 0 \\ 0 & 0 & 0 & 0 & 0 \end{pmatrix}$$

If we do not define γ, we can consider as colluded agents the ones who satisfy the first two thresholds.

Then, let $\mathcal{S} = \{1, 2, 3, 4\}$, we obtain the vector $\mathbf{v} = (0, 0, 0, 0, 1)^T$. Now we set the parameter δ to 0.2.

Following the first approach A, the updated trust matrix is computed as:

$$\tilde{\mathbf{T}}_{\mathbf{A}} = \begin{pmatrix} 0 & 0.1580 & 0.0002 & 0.0004 & 0.0960 \\ 0.1660 & 0 & 0.0004 & 0.0002 & 0.0980 \\ 0.0002 & 0.0006 & 0 & 0.1860 & 0.0020 \\ 0.0004 & 0.0002 & 0.1840 & 0 & 0.0040 \\ 0.8334 & 0.8412 & 0.8154 & 0.8134 & 0.8000 \end{pmatrix}$$

Finally, by adopting our new method B and by choosing $\varphi = 0.0004$, the trust matrix will be:

$$\tilde{\mathbf{T}}_{\mathbf{B}} = \begin{pmatrix} 0 & 0.0019 & 0.0124 & 0.0284 & 0.4800 \\ 0.0023 & 0 & 0.0249 & 0.0142 & 0.4900 \\ 0.0059 & 0.0143 & 0 & 0.0057 & 0.0100 \\ 0.0117 & 0.0048 & 0.0050 & 0 & 0.0200 \\ 0.9800 & 0.9791 & 0.9577 & 0.9517 & 0 \end{pmatrix}$$

In Fig. 4 the reputation vectors corresponding to cases A and B are plotted.

In this way, honest good agents were identified as colluded malicious, whereas the single honest agent a_5 has been given a very high reputation.

Now we study the IoT community correctly by using the complete threshold strategy that we described in Sect. 4. Therefore, let $\gamma = 0.25$, then we obtain $\mathcal{S} = \{3, 4\}$ and the vector $\mathbf{v} = (\frac{1}{3}, \frac{1}{3}, 0, 0, \frac{1}{3})^T$. By following the first approach A, the updated trust matrix is computed as:

$$\tilde{\mathbf{T}}_{\mathbf{A}} = \begin{pmatrix} 0.2667 & 0.4247 & 0.2669 & 0.2671 & 0.3627 \\ 0.4327 & 0.2667 & 0.2671 & 0.2669 & 0.3647 \\ 0.0002 & 0.0006 & 0 & 0.1860 & 0.0020 \\ 0.0004 & 0.0002 & 0.1840 & 0 & 0.0040 \\ 0.3001 & 0.3079 & 0.2821 & 0.2801 & 0.2667 \end{pmatrix}$$

Finally, by adopting our new method B and by choosing $\varphi = 0.0004$, the trust matrix will be:

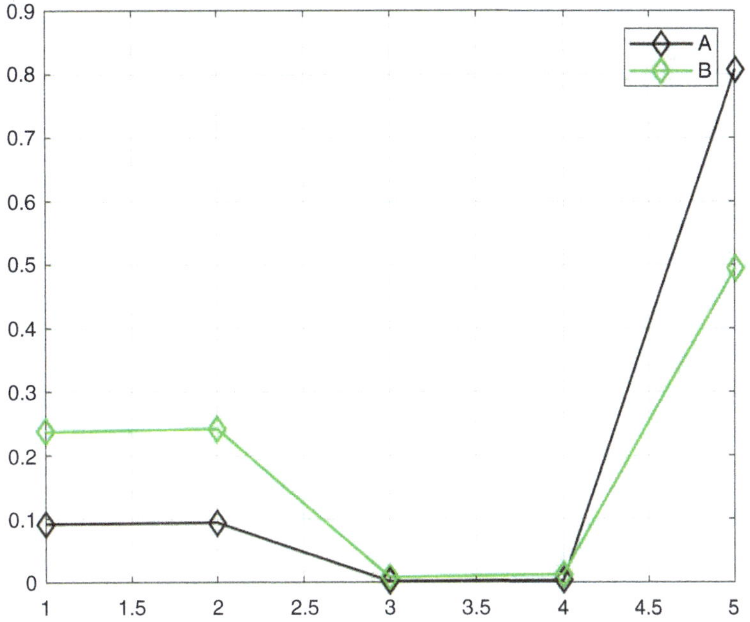

Fig. 4 The reputation vectors obtained with the approaches A and B

$$\tilde{\mathbf{T}}_{\mathbf{B}} = \begin{pmatrix} 0 & 0.7900 & 0.0124 & 0.0284 & 0.4800 \\ 0.8300 & 0 & 0.0249 & 0.0142 & 0.4900 \\ 0.0010 & 0.0030 & 0 & 0.0057 & 0.0100 \\ 0.0020 & 0.0010 & 0.0050 & 0 & 0.0200 \\ 0.1670 & 0.2060 & 0.9577 & 0.9517 & 0 \end{pmatrix}$$

In Fig. 5 the correct reputation vectors corresponding to cases A and B are plotted.

It is clear that γ threshold is fundamental for the correct analysis of the community. Moreover, also in this case, our method B returns reputation values consistent as much as possible with the true nature of the agents.

6 Conclusions

By interacting with the world around them, IoT objects can build multidimensional and context-sensitive smart network infrastructures potentially rich of social interactions. Such social activities can be boosted by exploiting the convergence between IoT and agent technologies. They enable scalable and pervasive cooperation and

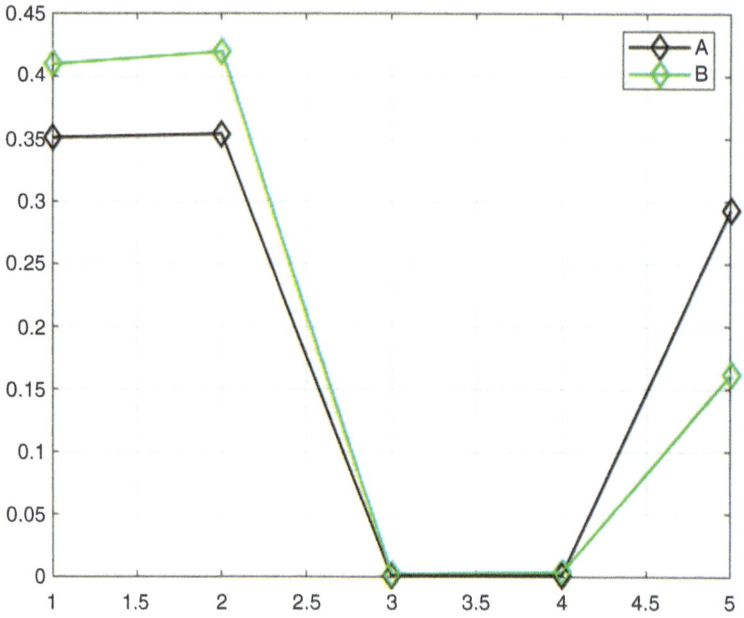

Fig. 5 The reputation vectors obtained with the approaches A and B

interactions activities, service discovery, and/or composition and allow IoT objects to overcome each form of heterogeneity.

In such an IoT scenario, the presence of malicious actors should be considered to minimize the risks of unsatisfactory interactions due to cheating activities performed by means of different malicious behaviors. Such risks can be increased in the presence of large IoT environments, nomadic IoT objects, and other events. Minimizing these risks is of primary importance to enhance both smart objects social aspects and user acceptance and consumption of IoT services and applications. To this purpose, an effective solution is to provide each player with appropriate trust measures about potential partners, and, in this respect, we presented: (*i*) A method to preliminarily identify the best candidates as malicious (colliding) in order to use such agents as pre-untrusted entities. (*ii*) A novel reputation model to detect collusive malicious agents that, differently from other TRSs, do not penalize reputation scores of honest agents in detecting malicious ones.

Finally, our forthcoming researches will be focused on (*i*) realizing an experimental campaign of tests on real and simulated data to confirm the good results obtained in our preliminary tests about both robustness against collusive attacks and correctness in computing honest actors' trustworthiness measures and (*ii*) then on considering also different attack modalities.

References

1. Wafa Abdelghani, Corinne Amel Zayani, Ikram Amous, and Florence Sèdes. Trust management in social internet of things: a survey. In *Conference on e-Business, e-Services and e-Society*, pages 430–441. Springer, 2016.
2. Anna A Adamopoulou and Andreas L Symeonidis. A simulation testbed for analyzing trust and reputation mechanisms in unreliable online markets. *Electronic Commerce Research and Applications*, 13(5):368–386, 2014.
3. Adnan Ahmed, Kamalrulnizam Abu Bakar, Muhammad Ibrahim Channa, Khalid Haseeb, and Abdul Waheed Khan. A survey on trust based detection and isolation of malicious nodes in ad-hoc and sensor networks. *Frontiers of Computer Science*, 9(2):280–296, 2015.
4. Ayesha Altaf, Haider Abbas, Faiza Iqbal, and Abdelouahid Derhab. Trust models of internet of smart things: A survey, open issues, and future directions. *Journal of Network and Computer Applications*, 137:93–111, 2019.
5. Parvaneh Asghari, Amir Masoud Rahmani, and Hamid Haj Seyyed Javadi. Service composition approaches in iot: A systematic review. *Journal of Network and Computer Applications*, 120:61–77, 2018.
6. Kevin Ashton. That' internet of things' thing. rfid journal, 22 june 2009, 2009.
7. Luigi Atzori, Antonio Iera, Giacomo Morabito, and Michele Nitti. The social internet of things (siot)–when social networks meet the internet of things: Concept, architecture and network characterization. *Computer networks*, 56(16):3594–3608, 2012.
8. Amir Jalaly Bidgoly and Behrouz Tork Ladani. Modelling and quantitative verification of reputation systems against malicious attackers. *The Computer Journal*, 58(10):2567–2582, 2015.
9. Amir Jalaly Bidgoly and Behrouz Tork Ladani. Benchmarking reputation systems: A quantitative verification approach. *Computers in Human Behavior*, 57:274–291, 2016.
10. Luis Cabral and Ali Hortacsu. The dynamics of seller reputation: Evidence from ebay. *The Journal of Industrial Economics*, 58(1):54–78, 2010.
11. Jin-Hee Cho, Kevin Chan, and Sibel Adali. A survey on trust modeling. *ACM Computing Surveys (CSUR)*, 48(2):1–40, 2015.
12. Pasquale De Meo, Lidia Fotia, Fabrizio Messina, Domenico Rosaci, and Giuseppe M. L. Sarné. Providing recommendations in social networks by integrating local and global reputation. *Information Systems*, 78:58–67, 2018.
13. Pasquale De Meo, Fabrizio Messina, Maria Nadia Postorino, Domenico Rosaci, and Giuseppe M. L. Sarné. A reputation framework to share resources into iot-based environments. In *IEEE 14th Int. Conf. on Networking, Sensing and Control*, pages 513–518. IEEE, 2017.
14. Ikram Ud Din, Mohsen Guizani, Byung-Seo Kim, Suhaidi Hassan, and Muhammad Khurram Khan. Trust management techniques for the internet of things: A survey. *IEEE Access*, 7:29763–29787, 2018.
15. Weidong Fang, Wuxiong Zhang, Wei Chen, Tao Pan, Yepeng Ni, and Yinxuan Yang. Trust-based attack and defense in wireless sensor networks: a survey. *Wireless Communications and Mobile Computing*, 2020, 2020.
16. Giancarlo Fortino, Lidia Fotia, Fabrizio Messina, Domenico Rosaci, and Giuseppe M. L. Sarné. Trust and reputation in the internet of things: state-of-the-art and research challenges. *IEEE Access*, 8:60117–60125, 2020.
17. Giancarlo Fortino, Fabrizio Messina, Domenico Rosaci, and Giuseppe M. L. Sarné. Using blockchain in a reputation-based model for grouping agents in the internet of things. *IEEE Transactions on Engineering Management*, 67(4):1231–1243, 2019.
18. Giancarlo Fortino, Fabrizio Messina, Domenico Rosaci, and Giuseppe M. L. Sarné. Resiot: An iot social framework resilient to malicious activities. *IEEE/CAA Journal of Automatica Sinica*, 7(5):1263–1278, 2020.

19. Karen K Fullam, Tomas B Klos, Guillaume Muller, Jordi Sabater, Andreas Schlosser, Zvi Topol, K Suzanne Barber, Jeffrey S Rosenschein, Laurent Vercouter, and Marco Voss. A specification of the agent reputation and trust (art) testbed: experimentation and competition for trust in agent societies. In *Proceedings of the fourth international joint conference on Autonomous agents and multiagent systems*, pages 512–518, 2005.
20. Diego Gambetta et al. Can we trust trust. *Trust: Making and breaking cooperative relations*, 13:213–237, 2000.
21. Seyed Asgary Ghasempouri and Behrouz Tork Ladani. Modeling trust and reputation systems in hostile environments. *Future Generation Computer Systems*, 99:571–592, 2019.
22. Stephen C Hayne, Haonan Wang, and Lu Wang. Modeling reputation as a time-series: Evaluating the risk of purchase decisions on ebay. *Decision Sciences*, 46(6):1077–1107, 2015.
23. Ferry Hendrikx, Kris Bubendorfer, and Ryan Chard. Reputation systems: A survey and taxonomy. *Journal of Parallel and Distributed Computing*, 75:184–197, 2015.
24. Kevin Hoffman, David Zage, and Cristina Nita-Rotaru. A survey of attack and defense techniques for reputation systems. *ACM Computing Surveys (CSUR)*, 42(1):1–31, 2009.
25. T.D. Huynh, N.R. Jennings, and N.R. Shadbolt. An integrated trust and reputation model for open multi-agent systems. *Autonomous Agents and Multi-Agent Systems*, 13(2):119–154, 2006.
26. HK Jnanamurthy and Sanjay Singh. Detection and filtering of collaborative malicious users in reputation system using quality repository approach. In *2013 International Conference on Advances in Computing, Communications and Informatics (ICACCI)*, pages 466–471. IEEE, 2013.
27. Audun Jøsang. Robustness of trust and reputation systems: Does it matter? In *IFIP International Conference on Trust Management*, pages 253–262. Springer, 2012.
28. Audun Jøsang and Jennifer Golbeck. Challenges for robust trust and reputation systems. In *Proceedings of the 5th International Workshop on Security and Trust Management (SMT 2009), Saint Malo, France*, volume 5. Citeseer, 2009.
29. Audun Jøsang, Roslan Ismail, and Colin Boyd. A survey of trust and reputation systems for online service provision. *Decision support systems*, 43(2):618–644, 2007.
30. S.D. Kamvar, M.T. Schlosser, and H. Garcia-Molina. The eigentrust algorithm for reputation management in P2P networks. In *Proc. of World Wide Web, 12th International Conference on*, pages 640–651. ACM, 2003.
31. Sepandar D Kamvar, Mario T Schlosser, and Hector Garcia-Molina. The eigentrust algorithm for reputation management in p2p networks. In *Proc. of the 12th international conference on World Wide Web*, pages 640–651. ACM, 2003.
32. Reid Kerr and Robin Cohen. Treet: The trust and reputation experimentation and evaluation testbed. *Electronic Commerce Research*, 10(3):271–290, 2010.
33. Gerd Kortuem, Fahim Kawsar, Vasughi Sundramoorthy, and Daniel Fitton. Smart objects as building blocks for the internet of things. *IEEE Internet Computing*, 14(1):44–51, 2009.
34. Gianluca Lax and Giuseppe M. L. Sarné. CellTrust: a reputation model for C2C commerce. *Electronic Commerce Research*, 8(4):193–216, 2006.
35. Seppo Leminen, Mika Westerlund, Mervi Rajahonka, and Riikka Siuruainen. Towards iot ecosystems and business models. In *Internet of things, smart spaces, and next generation networking*, pages 15–26. Springer, 2012.
36. Liang Liu, Zuchao Ma, and Weizhi Meng. Detection of multiple-mix-attack malicious nodes using perceptron-based trust in iot networks. *Future generation computer systems*, 101:865–879, 2019.
37. Somayya Madakam, Vihar Lake, Vihar Lake, Vihar Lake, et al. Internet of things (IoT): A literature review. *Journal of Computer and Communications*, 3(05):164, 2015.
38. C Marche and M Nitti. Trust-related attacks and their detection: a trust management model for the social IoT. *IEEE Transactions on Network and Service Management*, 2020.
39. Félix Gómez Mármol and Gregorio Martínez Pérez. Security threats scenarios in trust and reputation models for distributed systems. *computers & security*, 28(7):545–556, 2009.
40. Félix Gómez Mármol and Gregorio Martínez Pérez. TRMSim-WSN, trust and reputation models simulator for wireless sensor networks. In *2009 IEEE International Conference on Communications*, pages 1–5. IEEE, 2009.

41. D Harrison McKnight and Norman L Chervany. The meanings of trust. 1996.
42. Fabrizio Messina, Giuseppe Pappalardo, Domenico Rosaci, Corrado Santoro, and Giuseppe M. L. Sarné. A trust-aware, self-organizing system for large-scale federations of utility computing infrastructures. *Future Generation Computer Systems*, 2015.
43. Behrouz Pourghebleh, Karzan Wakil, and Nima Jafari Navimipour. A comprehensive study on the trust management techniques in the internet of things. *IEEE Internet of Things Journal*, 6(6):9326–9337, 2019.
44. P. Resnick, R. Zeckhauser, Friedman E., and K. Kuwabara. Reputation systems. *Communication of ACM*, 43(12):45–48, 2000.
45. Paul Resnick and Richard Zeckhauser. Trust among strangers in internet transactions: Empirical analysis of ebay's reputation system. In *The Economics of the Internet and E-commerce*. Emerald Group Publishing Limited, 2002.
46. Domenico Rosaci, Giuseppe M. L. Sarnè, and Salvatore Garruzzo. Integrating trust measures in multiagent systems. *International Journal of Intelligent Systems*, 27(1):1–15, 2012.
47. Syed Muhammad Sajjad, Safdar Hussain Bouk, and Muhammad Yousaf. Neighbor node trust based intrusion detection system for WSN. *Procedia Computer Science*, 63:183–188, 2015.
48. Amirali Salehi-Abari and Tony White. Dart: A distributed analysis of reputation and trust framework. *Computational Intelligence*, 28(4):642–682, 2012.
49. Claudio Savaglio, Giancarlo Fortino, and Mengchu Zhou. Towards interoperable, cognitive and autonomic iot systems: An agent-based approach. In *2016 IEEE 3rd World Forum on Internet of Things (WF-IoT)*, pages 58–63. IEEE, 2016.
50. Avani Sharma, Emmanuel S Pilli, Arka P Mazumdar, and Poonam Gera. Towards trustworthy internet of things: A survey on trust management applications and schemes. *Computer Communications*, 2020.
51. Wei She, Qi Liu, Zhao Tian, Jian-Sen Chen, Bo Wang, and Wei Liu. Blockchain trust model for malicious node detection in wireless sensor networks. *IEEE Access*, 7:38947–38956, 2019.
52. Sokratis Vavilis, Milan Petković, and Nicola Zannone. A reference model for reputation systems. *Decision Support Systems*, 61:147–154, 2014.
53. Tian Wang, Pan Wang, Shaobin Cai, Xi Zheng, Ying Ma, Weijia Jia, and Guojun Wang. Mobile edge-enabled trust evaluation for the internet of things. *Information Fusion*, 75:90–100, 2021.
54. Yu-Feng Wang, Yoshiaki Hori, and Kouichi Sakurai. Characterizing economic and social properties of trust and reputation systems in p2p environment. *Journal of Computer Science and Technology*, 23(1):129–140, 2008.
55. Li Xiong and Ling Liu. Peertrust: Supporting reputation-based trust for peer-to-peer electronic communities. *IEEE transactions on Knowledge and Data Engineering*, 16(7):843–857, 2004.
56. Zheng Yan, Peng Zhang, and Athanasios V Vasilakos. A survey on trust management for internet of things. *Journal of network and computer applications*, 42:120–134, 2014.

An Adaptive Blurring Routing Protocol for Delay-Tolerant Networks in IoT Environments

D. Meli, F. L. M. Milotta, C. Santoro, Federico Fausto Santoro, and S. Riccobene

1 Introduction

Delay-tolerant networks (DTNs) [5] are particular networks with the peculiar feature that it does not exist a complete or a defined path to the destination, with its links available at the same time. In such environments, which are often featured by IoT system, the routing protocol must exploit the node movements and meet to physically carry messages to their destinations. Consequently, to increase the assurance of delivery, the communication time constraints are scarified to find a complete end-to-end routing path. Therefore, movements and meetings among nodes are *opportunities* to find new partial paths towards destination. The "store, carry and forward" paradigm represents the main way to implements routing protocols on DTNs.

In [3] authors present a taxonomy on routing protocols for DTN, distinguishing between pure opportunistic approaches and social-based ones. All the protocols presented ground their functioning on the knowledge of interactions among nodes, in terms of strictly social behaviour or node movements and encounters. The main problem of all routing protocols is to find a correct next hop to reach the destination. In the specific case of DTN we have to decide if the encounter node can be a good carrier, how much it is "good", i.e. how many copies of the message we have to distribute and how much time a message must be stored, waiting for other possible good carrier. The knowledge of node behaviour is then fundamental to take correct decisions regarding message delivery. At the same time, this knowledge provides sensitive information on nodes, violating their privacy.

D. Meli · F. L. M. Milotta · C. Santoro · F. F. Santoro (✉) · S. Riccobene
Department of Mathematics and Computer Science, University of Catania, Catania, Italy
e-mail: danilo.meli@unict.it; milotta@dmi.unict.it; santoro@dmi.unict.it; federico.santoro@unict.it; riccobene@unict.it

L. Fotia et al. (eds.), *Security, Trust and Privacy Models, and Architectures in IoT Environments*, Internet of Things, https://doi.org/10.1007/978-3-031-21940-5_4

A simple approach to DTN routing problem is presented in [15]. The routing protocol, called *epidemic*, uses a modified flooding approach to deal with the path discovery. As the name suggests, a message is replicated in every new met node, so every possible route is tested. In terms of communication bandwidth and storage memory, this strategy clearly wastes a lot of resources. The number of the copies to distribute is limited by the memory buffer, handled with a FIFO policy.

Spray and Wait [16] is a modified version of epidemic routing protocol that tries to reduce communication overhead addressing messages only towards the supposed destination area. It works in two phases: in the first one, the so-called spray phase, a new created message is distributed in, at least, k copies to specific relay nodes, similarly to the epidemic strategy. After that, if the destination node is not reached yet, the *wait* (second) phase starts: no other copy will be generated, and every relay node will deliver the message only to the correct destination. These two examples do not violate the node privacy, as they do not use sensitive information. The flooding approach, used by both protocols, is essentially zero knowledge but wastes a lot of resources, as node memory and network messages.

Other more efficient DTN routing protocols base their decisions on past history: in real life the node movements are not random but follow a social behaviour. In these cases, there is a nontrivial probability that two specific nodes, which met each other in the past, will meet themselves again in the future.

PRoPHET [8] bases its routing protocol on the knowledge of these behaviours. When a node encounters another one, it calculates the probability the second one could reach directly or indirectly the final destination. To implement this approach, every node has to maintain a vector, contained the so called delivery predictability (DP), to keep trace of previous meeting, in order to calculate future meeting probability. The DP is evaluated with the following two formulae:

$$P_A(B)_{new} = P_A(B)_{old} + (1 - P_A(B)_{old}) \cdot P_{init} \tag{1}$$

used to update the DP from A to B:

$$P_A(C)_{new} = P_A(C)_{old} + (1 - P_A(C)_{old}) \cdot P_A(B) \cdot P_B(C) \cdot \beta \tag{2}$$

as a transitive rule, referred to a third node C.

All these DPs are then shared with other nodes at every encounter, in order to allow other DP update. The correctness of the DPs is the base for correct routing decisions.

The exchange of this information among nodes represents an obvious weakness in terms of *privacy preserving*. Malicious intruders can collect them to easily rebuild the graph of node contacts and movements, violating their privacy.

Privacy is defined by Clifton et al. [4] as "the prevention of personal data usage in a way that negatively impacts someone's life". Preserving privacy can therefore be referred to as a measure of preventing a malicious user to exploit information for identification of a physical being or discover relationships physical being. The main aim of this work is to design a routing protocol for opportunistic DTN, where sensitive information is preserved by means pseudo-anonymization.

The protocol draws inspiration from the human ability to adapt and get influenced by the environment, in terms of vocal tone and timbre of the speaking language. A node will be influenced by changing its place and by meets with new nodes and can address information with its and environment's characteristics. These characteristics are defined by a mathematical function that represents a node's track. The routing protocol realizes on the comparison of these tracks and on the evolution of the tracks shape, because these will evolve continuously in time and being more similar to the tracks of met nodes. The protocol is divided in phases. Firstly the node joins to the network and generates its track, and it does so by the near nodes. Secondly, when a meet occurs between nodes, their tracks will be changed with an evolution function. Finally, a node that wants address a message will decide if it will give the message to the met node with a matching function. Moreover, all meets with new nodes will change the track and will forget the acknowledgements about its past.

The chapter is structured as follows. Section 2 reports the state of the art in terms of protocols for delay-tolerant networks. Section 3 introduces the idea behind the proposed protocol. Section 4 presents the protocol description, with its phases and mathematical models. Section 5 discusses a case study and shows experimental data stating how the proposed protocol outperforms the classical DTNs routing protocols, in terms of delivery performances. Finally, the Sect. 6 draws our conclusions and outlines the future works.

2 Related Works

Nowadays, *privacy preserving* is assuming a role even more important inside digital communications because a lot of data, often related to our daily life, are conveyed into network infrastructures. For instance, several mobile applications share our personal data for job search or other private purposes, that can be spoofed for malicious intents. An example could also be the context of IoT (Internet of Things) [2]: it is composed by a large number of small devices, which share a big amount of data that may be also sensitive. In fact, with the increasing of the home-automation, many of these devices can be embedded in our home furnishings. In the context of IoT, DTNs represent a good choice to implement the network infrastructure.

The problem of privacy preserving in DTNs has been addressed following two ways: *(i)* by blurring data and *(ii)* by using an homomorphic encryption (HE) schemes. In [17] the authors describe an approach based on multiple paths for each destination to prevent traffic analysis. The proposed architecture uses layered encryption scheme based on a PKI. A message is split in parts using erasure code and each part follows a different path.

Bloom filters are another way to data blurring. In [11] two methods are presented, based on space-efficient probabilistic data structures, used to represent sets. The first one substantially reduces the representation space overhead using a summary vector. The second one uses a Bloom filter in opportunistic networks. When two nodes encounter each other, they exchange the information stored in their buffers to avoid useless transmissions and control redundancy, through a summary vector indicating the packets already received. However, we must underline that Bloom filters introduce some drawbacks like additional processing overhead and, principally, the presence of a "false positive". To reduce this phenomenon, Bloom filters should be used in a very large scale of node sets.

Another approach to the privacy preserving problem in DTN routing is grounded on the use of homomorphic encryption (HE) schemes [1, 6]. In [14] authors present a routing protocol based on the concept of habitat, defined as the main zone where a node stay in. To decide which node can be the best carrier towards the destination, the protocol calculates which, between two candidates, can provide a habitat closer to the destination one. This is done with the exchange of encrypted coordinates, representing a simplified version of the habitat zone. However, the protocol presents two main weaknesses: *(i)* The communication involves only one intermediate carrier: a chain of carrier is hard to find. *(ii)* The concept of habitat works well in case of long permanence of a node in a given limited zone.

Also other works are based on HE [9]: the main difficulty to apply this technique is the limited type of operations that HE permits.

3 Innovation from Natural Context and Proposed Solution: Privacy Preserving Delay-Tolerant Network (PPDTN)

When somebody learns to speak a language, she/he is influenced by the country where she/he is and by the people which live there. Changing place and/or meeting different people, he will also change his way to talk, in terms of tone, accent or syntax. In fact, when we listen to someone speaking around, we can guess where he comes from, with a good accuracy.

Let us now transpose this concept in a network domain, where we have DTNs instead of countries and nodes instead of people. A node can address a message using some peculiar characteristics of destination. These features are specific not only of the wanted node but also of the zone and the context where destination node currently is assimilated by it.

Some DTNs use opportunistic routing protocols to exchange messages, but the cost to implement a correct and efficient routing is the sharing of sensitive information on node movements and topology throughout the network. The solution presented in this chapter minimizes the usage of specific information linked to the nodes behaviours, preferring information related to the environment.

As in human context we speak about vocal tone or timbre [13], in a DTN environment, we will use a peculiar track for every node: each track shape will be moulded by meeting history. The track can be simply thought as a mathematical function. Routing will be based on the similarity of the node tracks. In [7] authors present a similar approach, based on so called "labels". A label indicates the hub affiliation of a node. The main difference with our approach is the evolution of the features (the track shape) that are bound to a node: a label is fixed for a hub; a track evolves with nodes encounters and is not linked only with a particular zone. In other words, with our protocol we relax the use of sensitive information from the encountered nodes, using instead more generic, and less sensitive, information.

As each person has a specific voice timbre, each node needs to have its own shape function that will be called "track". Track shapes are not permanent. A track will evolve continuously in time, trying to resemble to the ones of the met nodes.

If the tracks of two or more nodes are similar, it means they approximately go around to the same area and they know the same neighbours. However, note that this fact does not constitute a proof they met each other in the past.

4 Protocol Description

Let us introduce some features related to the new proposal:

- *Starting phase*: When a node joins to a network, it needs an initial track. This will be generated pseudo-randomly, for example, considering the ones of its neighbours.
- *Meeting phase*: When a node meets another one, their tracks must be accordingly modified in order to resemble each other. We need an *evolution function* that receives in input the two tracks (the main and the encountered track) and returns the modified track. This function will be applied by both nodes, separately.
- *Routing phase*: To address correctly a message, a node must decide how much the encountered node can be considered a good choice to reach a specific destination, as in any routing protocol based on opportunistic meeting. To implement this, we need a *matching function* that, receiving two tracks as input, returns an index of similarity, normalized between 0 and 1.
- *Other*: All the encounters modify a track accordingly to the evolution function. However, if a node changes for a long time its locations, it must forget past influences. Other new encounters will help this behaviour, but we need a faster method. The protocol provides an *ageing function* that modifies a track towards its initial shape. This represents a sort of forgetfulness procedure.

From a mathematical point of view, a track is a continuous function with some peculiar characteristics.

$$T(x) : \mathbb{R} \rightarrow \mathbb{R} \tag{3}$$

To easily maintain and handle it, we introduce the following characteristics:

- The function is defined in $[0, n]$ with $n \in \mathbb{N}$.

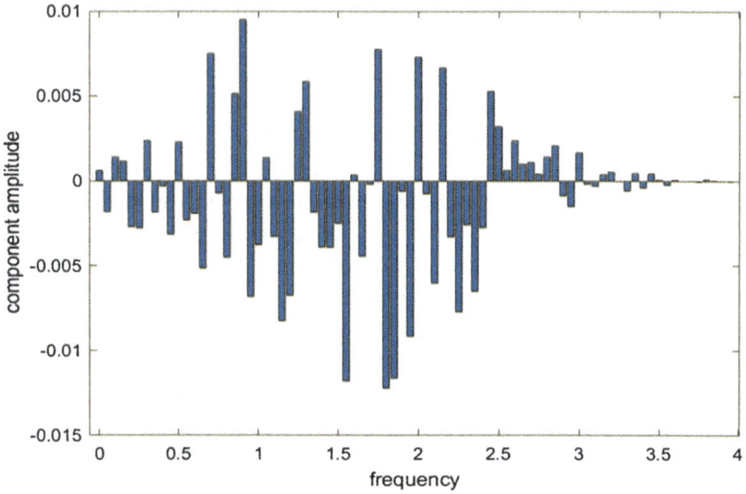

Fig. 1 Example of a node track, in frequency domain

- $T(x) \in [-m, m]$ with $m \in \mathbb{N}$.
- The function must have a strong variability in $[0, n]$.

The function becomes:

$$T(x) : [0, n] \rightarrow [-m, m] \tag{4}$$

All these requirements are introduced in order to correctly implement the proposed solution. The first two characteristics allow a simple comparison between two shapes. The last—the high variability in the defined interval $[0, n]$—is essential to strongly characterize the node and the context.

Starting Phase

When a node joins to the network, its track must be randomly generated. The high variability constrain can be obtained leveraging the frequency domain. If we randomly generate some frequency components—in the frequency domain—chosen in conformity to the definition interval, the desired function can be obtained by means the inverse Fourier transform.

Figures 1 and 2 present the spectrum in the frequency domain and the correspondent track in the time domain.

Store and handle a function of this kind can be very hard; so we sample the track in the interval of definition, choosing $n + 1$ equispaced value for the abscissa:

$$s_i = T(x_i) \qquad x_i = 0..n \tag{5}$$

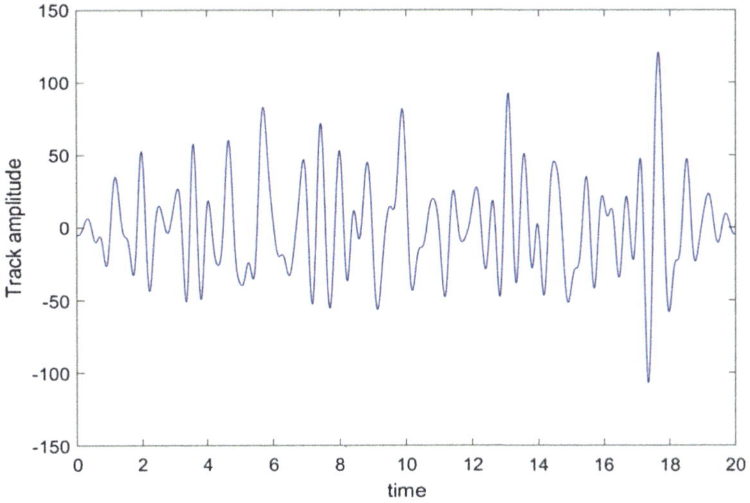

Fig. 2 Example of a node track, in time domain

obtaining $n + 1$ different points:

$$(0, s_0), (2, s_2), \ldots, \quad (n, s_n) \tag{6}$$

All the s_i values are then normalized and quantized into the amplitude $[-m, m]$. In the following, we will indicate them with y_i.

In this way, a track can be easily stored as a vector of $n + 1$ values. The set with all the points (x_i, y_i) identifies, with a high degree of accuracy, the original function.

Evolution Function

The track exchange during node encounters is a mandatory step to implement evolution and matching functions. To reinforce the privacy preserving property, it is useful to hide the exact function shape, sending instead a blurred representation obtained from the original one. To implement this, the protocol implies the following steps.

Let consider two numbers k and p such that:

$$k, p < n \quad k, p \in \mathbb{N}$$

$$n - 1 \leq p * k \leq n$$

We extrapolate a subset of k values from the original set:

$$x'_0 = x_0 \ldots x'_j = x_{j*p} \ldots x'_k = x_{k*p} \quad j \in [0, k]$$

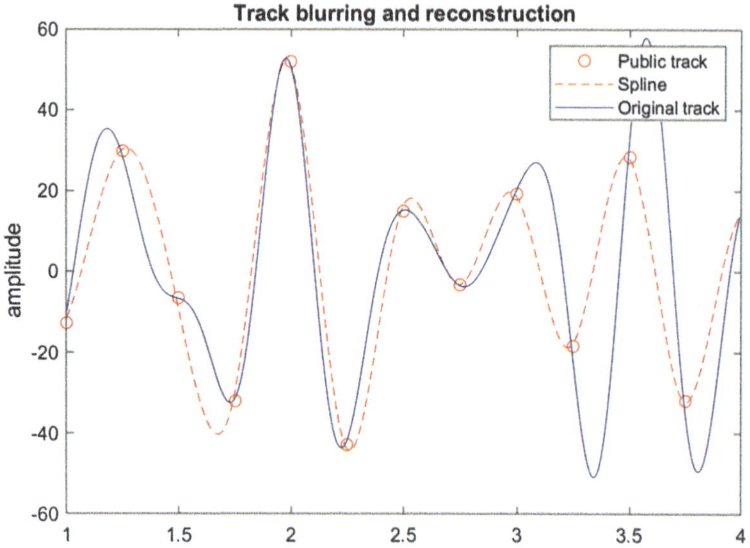

Fig. 3 A portion of the public track, the relative cubic spline interpolation and the original track

From these values we can derive the correspondent points on the function:

$$(x'_j, T(x'_j)) \quad j \in [0, k]$$

Note that the set $\{(x'_j, T(x'_j))\}$ represents an undersampling of the track.

After this step, the track is approximated with a cubic spline interpolation function [10, 12]; it is a 3-degree polynomial curve that can be derived twice throughout the definition interval. The blurred representation is what we call the public track of a node.

Figure 3 shows, in a zoomed view, the set of points of the public track, the correspondent cubic spline interpolation and the original track. Note that in some parts the approximation well fits the original track, but in other parts some peculiar characteristics are strongly blurred. The precision of the interpolation directly depends by the ratio k/n.

To implement the mutual influence between nodes in case of an encounter, we need an evolution function that, given a remote public track, modifies the real track of the node.

Let T_A be the real track of the node A and $T_{B_{public}}$ the public (blurred) track of B. We define:

$$f^{evol}(A, B, x) \stackrel{def}{=} \beta \cdot T_A(x) + (1 - \beta) \cdot T_{B_{public}}(x) \tag{7}$$

where $\beta \in]0.5, 1[$ is a coefficient that determines the impact of influence of remote tracks on track evolution.

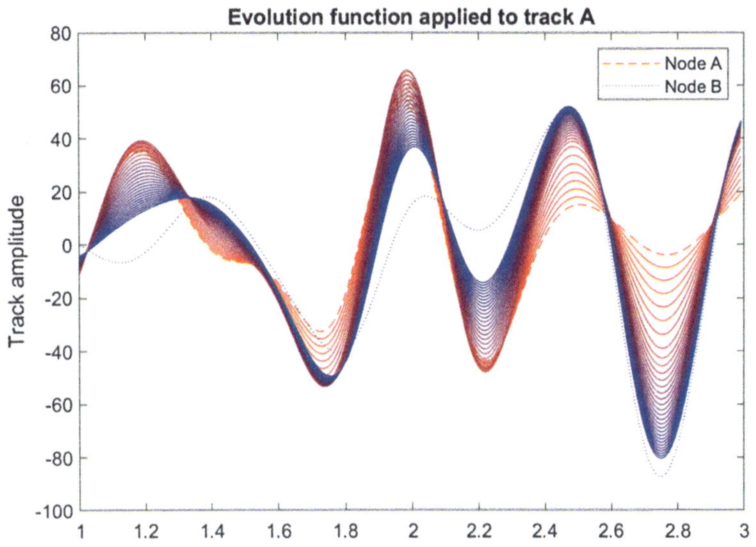

Fig. 4 Example of repeated applications of the evolution function to the node A track (zoomed view)

Node B communicates its track with a set of points:

$$T_{B_{public}}(x_j) \quad j \in [0, k] \tag{8}$$

Using these points, we derive the cubic spline that interpolates this set. Call it $S_B(x)$.

Applying the evolution function to node A track, will have:

$$T_A(x_i)_{new} = \beta \cdot T_A(x_i)_{old} + (1 - \beta) \cdot S_B(x_i) \quad i \in [0 \ldots n] \tag{9}$$

where $T_A(x)_{new}$ represents the evolved track.

It has been empirically observed that massive execution of the evolution function (9) tends to flatten the tracks, minimizing more and more the characteristics.

To avoid this behaviour, we apply the normalization procedure, as after the generation phase.

Figure 4 shows an example of the evolution for a track, applying 30 times the formula (9) and the normalization to the track of node A, keeping unmodified the other track (node B). The figure is a zoomed part of the entire shape. As we can see, when the two tracks present peaks in the same direction, as at time 1.8, the result accentuates the behaviour. Instead, when they are in opposition, the result tends to a mean value (see time 1.2). Further, the effect of the normalization procedure is visible in the movements of the peaks towards the ones of the track B (see, e.g., the interval between 1.0 and 1.6).

Ageing Function
As explained in Sect. 3, we need an ageing function that represents a sort of forgetfulness method, useful when a node changes its location and its contacts for a long time. This function modifies the track towards the original one. It can be implemented in a similar way of the evolution function and will be applied at constant time intervals.

$$f^{age}(A, x) \stackrel{def}{=} \gamma \cdot T_A(x) + (1 - \gamma) \cdot T_{A_{orig}}(x) \tag{10}$$

where $T_{A_{orig}}(x)$ is the original track of the node A, as generated in the starting phase, and γ is a parameter defined in]0, 1[.

Matching Function
The matching function provides a simple information on the ability of the encountered node to carry data towards the correct destination.

Starting from the profile of the destination and the public track of the met node, the function returns a value in the range [0, 1], where 1 means a perfect match.

Let $T_1(x)$ and $T_2(x)$ be the two functions, reconstructed with the cubic spline interpolation from the set of points associated with the destination and the met node. We define it in this way:

$$f^{match}(T_1, T_2) \stackrel{def}{=} 1 - \frac{\sum_{i=0}^{n} |T_1(i) - T_2(i)|}{2mn} \tag{11}$$

Every encountered node that determines a matching function result higher than a predetermined threshold can be considered a good carrier. Moreover, the set of results of the matching function for a given destination can be used to determine when to stop the research for other carrier, in order to consider the message as *delivered*, and so delete it.

5 Experimental Results

The proposed solution has been evaluated by means a simulator, with a scenario of 20×20 kilometres and a population of 100 nodes. The wireless communication range for the mobile nodes is about 250 m. The scenario has been divided in 16 zones; every node can travel along a limited number of contiguous zones, so there is a nontrivial possibility that 2 given nodes cannot encounter each other during the simulation. Also, we defined a set of meeting point: they represent the locations where the nodes move towards and stop for a certain time (600 s in average). The speed of the movements is variable between 3 and 30 m/s, similarly to human beings.

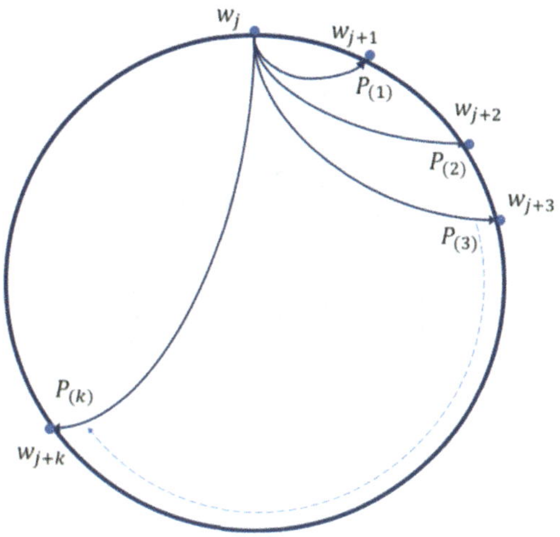

Fig. 5 Example of movements starting from the position w_j. The value $P(i)$ represents the probability that the next step is the point w_{j+i}

All these simulation parameters have been chosen to avoid that a given node could meet every other node but only a limited subset. The simulations highlight that this value is less than 15, with a mean value of 11 (approximately).

Every node follows a route that, with some probability, drives it towards a point chosen from the prefixed set of meeting points.

Let $L_i = \{w_1, w_2 \ldots w_{z_i}\}$ be the circular list of points chosen for node i and $z_i = |L_i|$.

When the node reaches the position w_j, we randomly select a value $k < z_i$ and a set of probability:

$p(x) : x \in [1, k] \mid \sum p(x) = 1.$

Every $p(x)$ value corresponds to a point in the list L_i that follows the point w_j.

In other words, the node follows a route that includes all the position of the list in a circular way, but some positions could be skipped in a given round. This behaviour reproduces the actual human one.

During the simulations, we monitored about 10,000 encounters (Fig. 5).

To validate our protocol, we compared its performance with the PRoPHET one. Figure 6 shows the distribution of the encounter probability obtained for all the occurrence of the encounters. For PRoPHET every bar represents the value that the encountered node communicates in order to reach the destination. For PPDTN, instead, it represents the results of the matching function. We compare these values each other because they are used to choose (or not) the encountered node as carrier. As we can see, both distributions present very similar shapes. In particular, the number of encounters with low value ($P < 0.70$) is high for both protocols, representing, respectively, the 94% and the 88% of the total.

To evaluate the actual working phase, we must extrapolate only the portion in the range 0.70 – 1.0 from this chart, because lower values are discarded by both protocols (i.e. the encountered node is judged as a useless carrier).

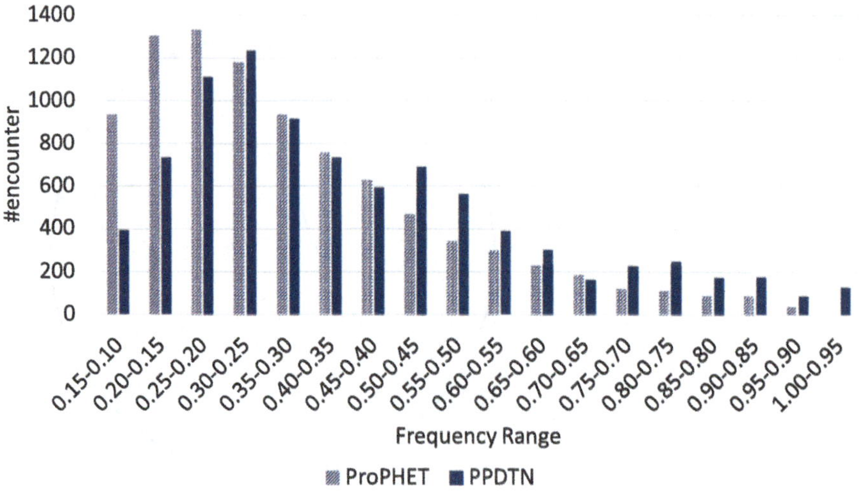

Fig. 6 Comparison between PRoPHET vs. PPDTN in terms of encounters that can reach a given target

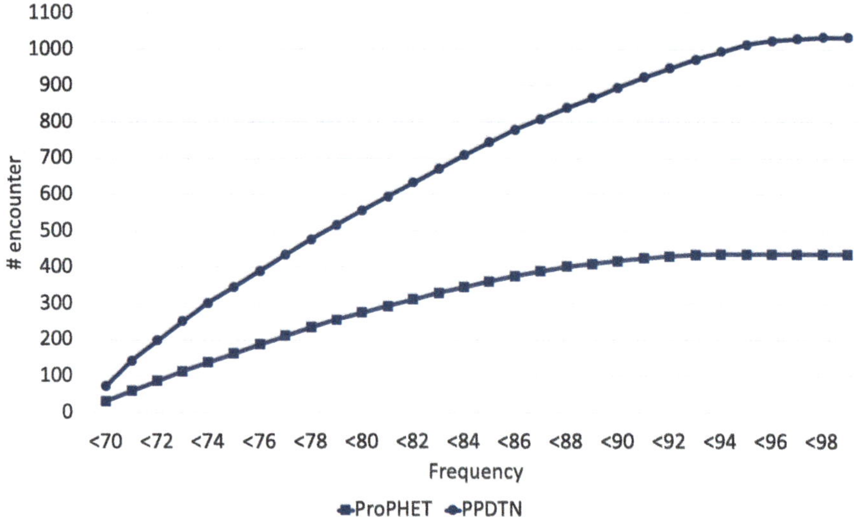

Fig. 7 Performances comparison between PRoPHET vs. PPDTN in the range of working value (values less than 0.70 have been erased)

In Fig. 7 the focus is placed in the range 0.70 – 1.0, considering only the encounters that provide a useful possibility for the delivery, higher than a given threshold. Every point indicates the number of useful encounters with a given probability. For example, we have about 480 encounters for PPDTN and 220 for PRoPHET with probability in the range 0.70, 0.78. As we can see, PRoPHET shows a limited increment over the value 0.86, forcing to choose a general threshold lower

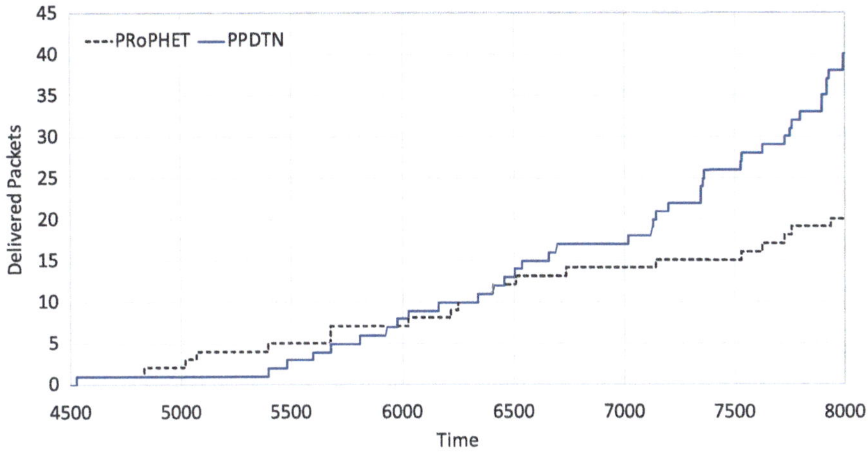

Fig. 8 Performances comparison between PRoPHET vs. PPDTN in the range of working value

than 0.80, in order to increment the number of nodes classified as "good choice". Instead PPDTN shows a constant increasing trend from 0.70 to 0.94, allowing a more punctual evaluation of the correct threshold. This last consideration can be used to improve the routing when the message is closer to the destination.

Within the simulation we also tested the message delivery performance obtained applying the rules defined for both protocols. After an initial running phase to reach the steady state, the simulator generates a message every 100 s on average, with random source and destination. The messages are not duplicated: when a node transfers a message to another one, the old copy will be deleted. Figure 8 diagrams the number of messages delivered for PRoPHET and PPDTN.

As we can see, only a small portion of the generated messages reaches its destination. This is in accord with the chosen scenario, with a large dimension (400 km^2) and a limited number of node (100). This was done to stress the protocols, highlighting flaws in the routing decisions.

The results show that PPDTN performance tends to double the PRoPHET one.

6 Conclusions

In IoT systems, routing protocols for DTNs are often based on the sharing of sensitive information, like movements and past encounters. This can lead to a non-collaborative behaviour from some nodes that see a lack of privacy in these protocols. In this paper we presented a new approach to the routing protocol that enforces the privacy preserving by means the sharing of generic and blurred information. The simulations carried out show good performance in reaching the correct destination.

References

1. Abbas, A., Hidayet, A., A.Selcuk, U., Conti, M.: A survey on homomorphic encryption schemes: Theory and implementation. ACM Computing Surveys **51**(4) (2018)
2. Benhamida, F.Z., Bouabdellah, A., Challal, Y.: Using delay tolerant network for the internet of things: Opportunities and challenges. In: 2017 8th International Conference on Information and Communication Systems (ICICS). pp. 252–257 (2017). https://doi.org/10.1109/IACS.2017.7921980
3. CC, S., Raychoudhury, V., Marfia, G., Singla, A.: A survey of routing and data dissemination in delay tolerant networks. Journal of Network and Computer Applications **67**, 128–146 (2016). https://doi.org/10.1016/j.jnca.2016.01.002, http://www.sciencedirect.com/science/article/pii/S1084804516000035
4. Clifton, C., Kantarcioglu, M., Vaidya, J.: Defining privacy for data mining. In: National science foundation workshop on next generation data mining. vol. 1, p. 1. Citeseer (2002)
5. Fall, K.: A delay-tolerant network architecture for challenged internets. In: Proceedings of the 2003 conference on Applications, technologies, architectures, and protocols for computer communications. pp. 27–34. ACM (2003)
6. Gentry, C., et al.: A fully homomorphic encryption scheme, vol. 20. Stanford university Stanford (2009)
7. Hui, P., Crowcroft, J., Yoneki, E.: Bubble rap: Social-based forwarding in delay-tolerant networks. IEEE Transactions on Mobile Computing **10**(11), 1576–1589 (2010)
8. Lindgren, A., Doria, A., Schelén, O.: Probabilistic routing in intermittently connected networks. In: ACM International Symposium on Mobilde Ad Hoc Networking and Computing, MobiHoc 2003: 01/06/2003-03/06/2003 (2003)
9. Magaia, N., Borrego, C., Pereira, P., Correia, M.: Privo: A privacy-preserving opportunistic routing protocol for delay tolerant networks. In: 2017 IFIP Networking Conference (IFIP Networking) and Workshops. pp. 1–9 (2017). https://doi.org/0.23919/IFIPNetworking.2017.8264835
10. McKinley, S., Levine, M.: Cubic spline interpolation. College of the Redwoods **45**(1), 1049–1060 (1998)
11. Papapetrou, E., Bourgos, V.F., Voyiatzis, A.G.: Privacy-preserving routing in delay tolerant networks based on bloom filters. In: 2015 IEEE 16th International Symposium on A World of Wireless, Mobile and Multimedia Networks (WoWMoM). pp. 1–9. IEEE (2015)
12. Quarteroni, A., Sacco, R., Saleri, F.: Numerical mathematics, vol. 37. Springer Science & Business Media (2010)
13. Schafer, R.W., Rabiner, L.R.: Digital representations of speech signals. Proceedings of the IEEE **63**(4), 662–677 (1975)
14. Sánchez-Carmona, A., Robles, S., Borrego Iglesias, C.: Privhab+: a secure geographic routing protocol for dtn. Computer Communications **78** (11 2015). https://doi.org/10.1016/j.comcom.2015.10.002
15. Vahdat, A., Becker, D., et al.: Epidemic routing for partially connected ad hoc networks (2000)
16. Wood, L., McKim, J., Eddy, W., Ivancic, W., Jackson, C.: Saratoga: A scalable file transfer protocol. work in progress as an Internet-draft (2010)
17. Zhu, Y., Hu, Y.: Making peer-to-peer anonymous routing resilient to failures. In: 2007 IEEE International Parallel and Distributed Processing Symposium. pp. 1–10. IEEE (2007)

Modeling and Detection of Denial-of-Sleep Attacks on Autonomous IoT Devices in Wireless Sensor Networks

Vasily Desnitsky

1 Introduction

Currently, wireless sensor networks are becoming more widespread in various application areas, namely, for monitoring the physical characteristics of the environment, premises, technical devices, and objects, inside vehicles and systems of connected manned and unmanned cars, and for remote monitoring of vital health characteristics of a human, animal, etc. Due to the dynamic spontaneous nature of the operation of such systems and changes in the location of wireless network nodes and communication modes, often the nodes of such networks often operate from autonomous power supplies. Unlike other types of impacts, a direct or indirect attack on a device's battery is, firstly, difficult to detect and, secondly, can be extremely effective. The difficulty in detecting such attacks stems from the fact that the charge consumption is often nonlinear, multifactorial, and poorly predictable. It is also determined by the specificity of such an attack as well as the lack of combined monitoring tools that involve the analysis of charge consumption, on the one side, and unauthorized actions of users and the behavior of other devices, on the other.

Depending on the specificity of the target device and, in particular, on how much the power consumption process is optimized by the device developers, canceling or preventing an attacker from such optimizations will increase battery consumption to one degree or another. In a number of scenarios, the execution of a Denial-of-Sleep attack can lead to the need for earlier maintenance of a stand-alone wireless

V. Desnitsky (✉)
St. Petersburg Federal Research Center of the Russian Academy of Sciences (SPC RAS), St. Petersburg, Russia
e-mail: desnitsky@comsec.spb.ru

L. Fotia et al. (eds.), *Security, Trust and Privacy Models, and Architectures in IoT Environments*, Internet of Things, https://doi.org/10.1007/978-3-031-21940-5_5

sensor network (WSN) node, which can be associated with significant financial losses. In addition, until such a device is properly maintained by personnel, the device is likely to be turned off and unable to perform its intended functions. In other scenarios, such as a flying drone, a Denial-of-Sleep attack could result in the drone crashing, being lost, or hijacked by an attacker [5]. All this determines the necessity and practical importance of developing tools for detecting Denial-of-Sleep attacks with high values of detection quality indicators. The most important indicators include a pair of precision and recall, as well as an f1-measure that integrally evaluates the detection quality. It is advisable to develop software tools for detecting Denial-of-Sleep attacks with the maximization of the above indicators. It is expedient in the conditions of different time-frequency characteristics of a Denial-of-Sleep attack and the effectiveness of various attack implementations.

The main contribution of the chapter lies in the combined software/hardware and simulation modeling of Denial-of-Sleep attacks targeted on autonomous nodes of wireless sensor networks. The modeling relies on a specific scenario with a two-stage attack initiated by an intruder that indirectly through legitimate nodes influence the victim node through legitimate nodes to prevent it from periodic passing to the sleep energy-efficient mode, thereby to deplete its energy resources and make it disabled for some time. The complexity of such attack detection is that the attacker disguises the messages as those of other legitimate nodes. Moreover, in general case the internal structure of such attacking messages is completely similar to benign ones, so the only possibility to catch them is not to work with individual attacking messages but to identify sequences of such messages constituting the attack. Therefore, the chapter contributes a machine learning-based approach to detect Denial-of-Sleep attacks in a quite precise way. The source data was modeled as logs extracted from network traffic recorded for normal communication and five variants of Denial-of-Sleep attacks. Testing on independent data samples confirmed quite high detection quality of the constructed solution. Possible enhancement of the approach and the implemented models is disclosed as future directions of the study.

The elements of the novelty of the obtained results include the hybrid modeling of Denial-of-Seep attacks, involving hardware assemble of WSN hardware, development of the normal functioning, and communication of the nodes and simulation modeling of attacks with various intensity modes. Besides, the novelty lies in a particular intelligent data analysis process driven for specific attacks and their features available for the analysis.

The rest of the chapter is organized as follows. Section 2 comprises a survey of existing literature in the field. Section 3 exposes an application scenario used for modeling Denial-of-Sleep attacks presented in Sect. 4. Section 5 contains description of the proposed machine learning-based detection mechanism and conducted experiments. Section 6 presents a discussion on the obtained results and Sect. 7 concludes the chapter.

2 Related Work

Currently, power depletion attacks and Denial-of-Sleep attacks in particular are increasingly applicable to stand-alone cyber-physical devices. A key condition for the applicability of such attacks is the use of depletable batteries inside the device as the main or additional source of energy [6, 8]. More specifically, as a separate subclass of devices, it is worth highlighting such devices that are able to support operation in two modes: firstly, the full-fledged operation mode and, secondly, the power-saving mode (sleep mode). In order to optimize the power consumption of such a device in the sleep mode, for a certain period of time, it limits or stops its communication and computing functions for a certain period of time and falls asleep until a preset or dynamically determined expected wake-up time [9]. The goal of an attacker attempting to perform a Denial-of-Sleep attack is to disrupt the normal process of cycling transfer of the device to and from sleep mode and hence deplete the accumulated battery power to a large extent.

First of all, such an attack may lead to a temporary inoperability of the device [4]. In this case, the resumption of operation of the device may be associated with its manual maintenance by an engineer, which may require significant time and possible financial and operating costs [2]. In addition, in case of critical cyber-physical systems, the consequences of Denial-of-Sleep attacks may include physical damage; loss of the attacked device, such as in the case of UAV; or negative impacts on the infrastructure of the device. The most crucial characteristics that are important both for the attacker and for detecting such an attack are, firstly, the effectiveness of the attack, expressed as the amount of reduction in battery power, and, secondly, the stealth of the attack, i.e., how difficult it might be to detect such an attack and how accurately this detection can be performed.

While the problems of analyzing and modeling Denial-of-Sleep attacks in wireless sensor networks are presented in published works quite widely [3, 4, 12, 13], issues of detecting such attacks have been becoming the subject of research in the last few years.

Nishanth and Mujeeb focus on SYN flooding-based Denial-of-Service attacks in wireless sensor networks and their detection [10]. The detection tool developed by these authors is oriented to use in a WSN in conditions of criticality of energy resources, and, as they state, it does not violate the principles of periodic transition of nodes to sleep mode. This work allows us to confirm the need for means of detecting various types of attacks in the WSN, which would imply efficient utilization and would not have Denial-of-Sleep as their side effect. The need to achieve energy efficiency of detection tools for Denial-of-Sleep, forge and replay attacks, as well as other attacks in a WSN is substantiated also in [8].

Several papers have been published on the detection of Denial-of-Sleep attacks in the context of particular domains of WSN applications. Razaque, et al. proposed a method for detecting collision, Denial-of Sleep, and selfish attacks associated with exploitation of energy consumption by the attacker at the medium access control level [11]. Dhunna and Al-Anbagi presented a low power cybersecurity mechanism for WSN-based smart grid monitoring applications [8] to address smart grid applications such as remote monitoring, equipment fault diagnostic, wireless advanced metering infrastructure, and residential energy management.

Delicato et al. proposed an approach based on the human immune system metaphor to detect some types of Denial-of-Sleep in the WSN by monitoring neighboring nodes and collaboratively identifying the intruder [4]. Bhattasali and Chaki disclosed an approach to detecting Denial-of-Sleep attacks in WSN based on the probabilistic absorbing Markov Chain [2]. In this case, such detection is based on the estimation of expected death time of the sensor network under normal scenario. The fact that an attack was detected is inferred from the disabling of the node, which, in turn, stems from comparison of its lifetime with the expected lifetime under conditions of its normal operation. Mohd et al. examine an approach to detecting Denial-of-Sleep attacks by using a support vector machine method and network simulation OPNET modeler that was used to generate the initial data [9].

The distinctive features of this work include the focus of the constructed Denial-of-Sleep attack detection model on the possibility of revealing such attacks at their early stages. Basing on current communications, it is feasible to detect a Denial-of-Sleep attack with high accuracy even before it is fully executed. Thus, it becomes possible to attempt minimizing the negative consequences of a Denial-of-Sleep attack both by preventing its successful completion and minimizing the infrastructural and financial damage from it. In addition, as part of the experimental study, based on the physical implementation of the ZigBee network, we proposed the use of some supervised machine learning models. Besides we identified specific means that allow detection with the best quality in conditions of limited initial data.

3 Application Scenario

To form an initial data for building attack detection tools, the following applied scenario for the operation of a self-organizing wireless sensor network was used. This scenario is based on ZigBee technology implementing a mesh topology network with Digi XBee version 2 nodes. The nodes of such networks dynamically form a network topology that changes over time with the roles of nodes, namely, coordinator, routers, and end nodes. In this case, the role of each node is set as the parameters of its firmware. Each of the network nodes, regardless of the role, includes an Arduino Uno R3/Mega 2560 microcontroller, an adapter board and a Digi XBee s2 wireless communication interface. Figure 1 shows a fragment of the assembled network, where there is an attack victim node, i.e., an autonomously operating mobile node with an end device role, several legitimately functioning nodes, as well as an attacker node, which performs a Denial-of-Sleep type of action involving legitimate nodes in the process performing an attack.

According to business goals and operating conditions, the victim node is responsible for episodic collection of primary data from its sensors and sending it to the network of other nodes for subsequent analysis of this data. Due to the

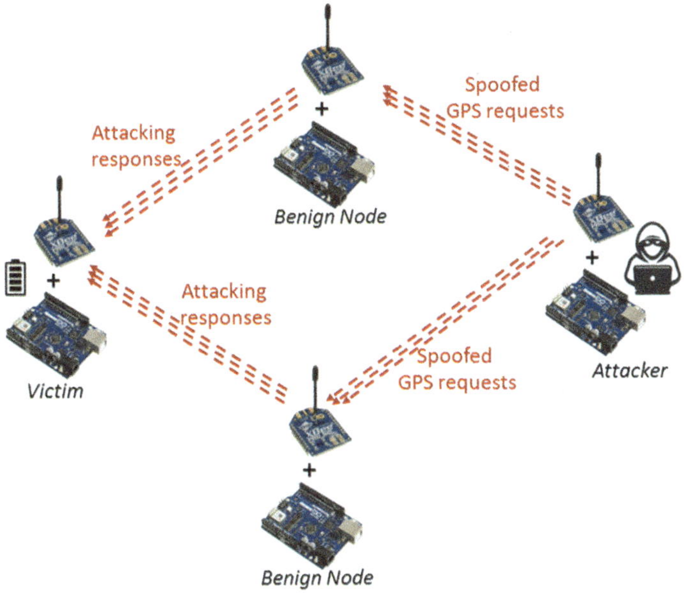

Fig. 1 A fragment of a wireless sensor network for modeling a Denial-of-Sleep attack

fact that this node can move in space, it supports the main, i.e., autonomous mode of operation, which is powered by the electric power storage. It should be noted that, in compliance with business needs, the collection and transfer of application data are performed at certain time intervals. Therefore, in order to ensure the energy efficiency of this node, its operation is organized by its cyclic transition from the full-fledged operation mode to the energy-efficient one (i.e., sleep mode). In the sleep mode, all the main energy intensive functions of the device are turned off, and the device waits for the onset of the wake time or an event which makes the device to wake up, e.g., the receipt of an incoming message. A Denial-of-Sleep attack in this scenario is a malicious exploitation of this mode-switching mechanism. As a part of such attack, messages are sent to the node at certain points in time, preventing the transition of this node from sleeping and thereby rapidly reducing its battery reserves.

Elements of full-scale software/hardware and simulation modeling were used to obtain initial data for building means for detecting Denial-of-Sleep attacks. The modeling was applied using the hardware of Arduino microcontrollers, XBee wireless communication interfaces, other electronic elements, and a personal computer necessary for microcontroller firmware, for more convenient reading of data from devices and testing of the entire network. The advantages of such modeling for the examined scenario include the proximity of such models of network nodes to the real conditions of WSN software/hardware operation.

Simulation modeling was used to increase the uniformity and simplify the modeling of normal communication processes in the network. Besides, it was applied in order to set Denial-of-Sleep attack vectors within modeling the duration of the attack, its intensity, and the structure of attacking messages (ZigBee protocol frames). The advantages of such modeling include the possibility and convenience of implementing difficultly reproducible scenarios for the operation of network nodes, which would be much more difficult to replay on a really working physical implementation. At the same time, such modeling takes into account the factors of randomness and the spontaneity of sending messages by nodes. Among other things these ones are determined by nondeterministic factors of the physical environment of the WSN and end user actions.

In general, a Denial-of-Sleep attack can be revealed in a coordinated set of actions by some malicious group of network nodes. Compared to a case of a single attacking node, such attack seems to be stealthier. It is far more difficult to trace the source of the attack when the message flow can include both messages that are actually related to the attack and messages from legitimate network nodes. However, without loss of generality, in order to implement attack detection functions within the framework of the simulation modeling, all malicious activity was modeled as using a single attacking node. But, to increase the stealth of such influence, the attacker can send attack commands to the victim node not explicitly but indirectly, forcing other legitimate network nodes to initiate attack traffic and send it to the victim node.

As a stream of attacking messages, some system and application commands were generated, and they were sent to the victim node and/or requested some information from it. A typical node vulnerable to this type of attack has the following power consumption mode. In accordance with the peculiarities of ZigBee technology, after receiving any message from the router node or after sending its own message, it waits for a certain specified time interval ST (time before sleep). After this time gap, in the absence of any network activity, it goes into sleep mode per time interval SP (cyclic sleep period) [6]. In general case, a Denial-of-Sleep attack is aimed at sending arbitrary or even empty messages to the node in the ST time interval so that the current ST report timer is reset to zero. And when getting to zero, the device switches to one of two states, namely, the communication state or standby state (without the possibility of switching to sleep mode). Moreover, as a rule, an intruder needs to perform such an attack most effectively to maximize the effect of battery discharge and the stealth of the attack for network detection tools, particularly in conditions of resource consumption restrictions of the attacker. This is formulated as follows:

$$\begin{cases} exhaustion(node_{victim}, attack) \underset{A}{\rightarrow} max \\ secrecy(attack) \underset{A}{\rightarrow} max \\ resources(attack) \leq r_0 \end{cases}$$

where $node_{victim}$ is the victim node, which power consumption is estimated by the exhaustion indicator. Values of the indicator are maximized on the set of possible implementations of Denial-of-Sleep attacks A. The secrecy indicator determines the level of stealth of the attack; r_0 is a constant of the restriction on resources available to the attacker.

During the normal operation of the network in the absence of these attack commands, such a command, for example, a request to the current GPS coordinates of a node, can be executed periodically by any node in the network to maintain up-to-date information about the network configuration that changes dynamically. The main difficulty in detecting a Denial-of-Sleep attack is to distinguish attacking messages from normal ones with maximum efficiency.

Figure 2 exposes a fragment of an example of the initial data received by the victim node and read for the development of tools for detecting a Denial-of-Sleep attack. This fragment of the logs of incoming messages is shown in the absence of an attack. The numerical values are presented in milliseconds. The data represents an enumeration of the absolute (left column) and normalized (right column) values relative to the start time, i.e., the time points of messages arriving at the attacked node. The red frames indicate the intervals of wakefulness between which the node can switch to sleep mode.

4 Attack Modeling

In this work, the logs of incoming traffic were recorded at the considered WSN node. Firstly, we fulfilled full-scale software/hardware modeling and recording of normal traffic reflecting regular communication in the network. This communication came as the result of simulation modeling of activity on the part of the third network node with low intensity, which was episodically sending requests to the analyzed node. Secondly, five simulation models of the attacker were built, differing in the degree of stealth of the attack. As with benign network activity modeling, each attack traffic entry contains timestamps within an approximately 10 min time interval.

The simulation modeling of normal activity in the network is performed by bunches, i.e., some set of messages (number of messages in a bunch) with a given interval both inside the bunch (interval inside bunches) and between the bunches themselves (interval between bunches) [7]. To enhance naturalness, when generating these parameters, additional allowable deviations are introduced. These deviations are corrected predeterminedly and/or pseudorandomly. In such simulation, these pseudorandom variables are distributed uniformly. Thus, several types of Denial-of-Sleep attacks were modeled not only with different degree of secrecy but with diversified attack intensity. Values of intensity were denoted by *high* and *low*.

Hence, for the Interval between bunches parameter, the values *high* = 20000 ms and *low* = 5000 ms were chosen with the maximum allowable deviations of 2000 ms and 500 ms, respectively. For the number of messages in a bunch parameter, we chose the values *high* = 20 pcs. and *low* = 5 pcs. with maximum allowable deviations

Fig. 2 A fragment of an example of initial data logs of messages received (in milliseconds)

1547725886971	0
1547725887320	349
1547725887854	883
1547725888230	1259
1547725888728	1757
1547725894047	7076
1547725894389	7418
1547725894918	7947
1547725895279	8308
1547725895809	8838
1547725896169	9198
1547725896684	9713
1547725902036	15065
1547725902894	15923
1547725903250	16279
1547725903749	16778
1547725904108	17137
1547725904452	17481
1547725909442	22471
1547725910194	23223
1547725910710	23739
1547725911079	24108
1547725911408	24437
1547725911941	24970

8 pcs. and 2 pcs., respectively. For the interval inside bunches parameter, the values *high* = 2000 ms and *low* = 500 ms were chosen with maximum tolerances of 667 ms and 167 ms, respectively.

Thus, the goal of the attacker is to prevent the victim node from going into sleep mode in order to increase its energy consumption. The following assumptions are accepted:

(1) The attacker is aware that the attacked node uses an XBee communication interface and communicates in API mode, which allows sending both broadcast messages and messages to specific recipients.
(2) The attacker is familiar with the structure of GPS coordinate request messages and knows the physical addresses of victim modules.
(3) The attacker managed to gain access to the software interface of one of the legitimate network nodes, namely, to the digital input of the node's hardware

platform, or introduced a false node completely controlled by the attacker into the network.

The following scenario for modeling Denial-of-Sleep attacks is implemented in the work:

(1) The attacker possesses a parasitic device that can connect to the serial interface of the module being exploited.
(2) The parasite starts generating data packets with a given frequency. The packets are similar to requests for GPS coordinates, indicating the address of the victim node in the sender data. In the experiment, this happens automatically, and the interval is adjusted with a certain step (e.g., 100 ms). The parasite also has a user interface consisting from a screen and buttons for entering the interval parameter.
(3) The attacker-operated module, which the parasite is connected to, receives these requests and, in accordance with its normal mode of operation, sends packets with coordinates to the victim node. This allows to carry out the attack with packets that do not have an anomalous structure.
(4) The victim node accepts both packets generated by the node in a normal cycle and those that were provoked by the parasite module. With a fairly frequent interval between packets, the ST parameter is reset more and more often, and the module stops falling asleep and starts consuming far more energy.

Let us illustrate the modeling described above with a series of diagrams visualizing the intensity of normal and attack traffic. In fact, the traffic was recorded on the victim node. Figure 3 exposes measurements for eight variants of triples of simulation modeling parameters, namely, <number of messages in a bunch, interval inside bunches, and interval between bunches>. For each of the parameters, symbols L and H correspond to *low* and *high* values, respectively, as described above. The

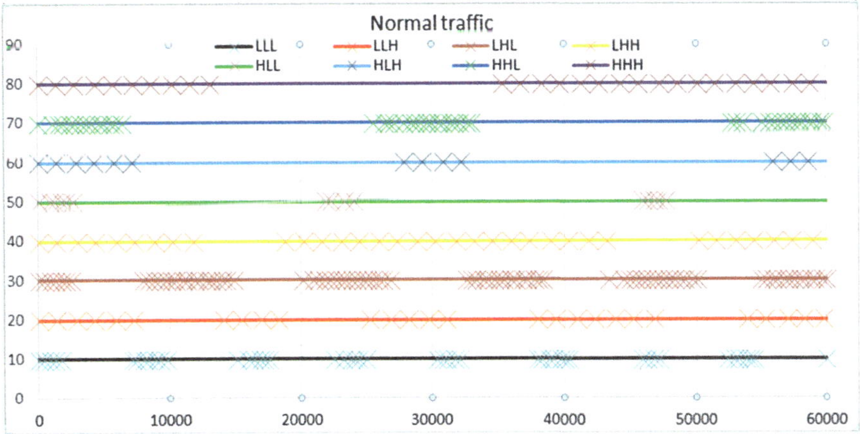

Fig. 3 Visualization of normal traffic

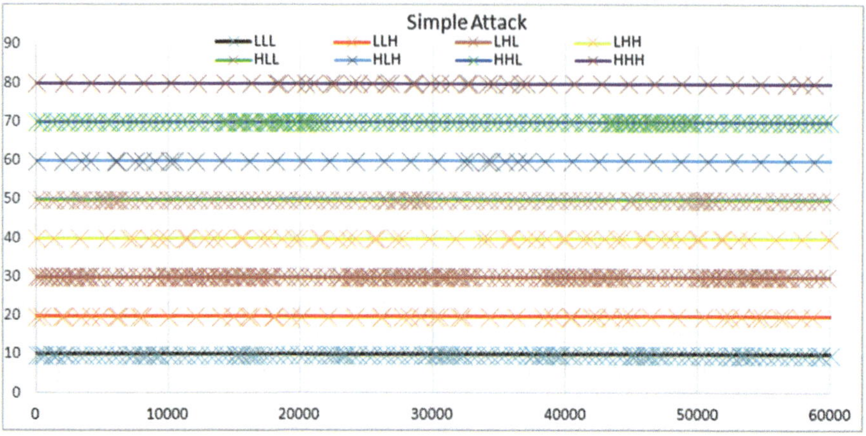

Fig. 4 Attack traffic visualization—simple attack

diagram depicts the first minute of traffic. It clearly shows the density of messages in a bunch. And one can also check the compliance with the intervals and the presence of deviations in the generation and transmission of messages. Each cross represents a fact that the message has arrived at the node. In addition, this diagram allows us to establish that the normal traffic generation parameters used, indeed, can significantly reduce power consumption due to the sleep mode that is activated between bunches of messages.

Let us consider five variations of the Denial-of-Sleep attack.

(1) Denial-of-Sleep attack variation 1 (simple attack)

The attack is performed with a frequency similar to the period of sending messages in the session. Firstly, it allows the intruder to maximally disguise his actions as real traffic. Secondly, there is no need to increase frequency of sending messages since the sleep mode parameters on the network should already be adjusted so that the end device does not fall asleep while the communication session is in progress. The results of the attack are shown in Fig. 4; the attack traffic is visible along the horizontal axes of normal traffic as in Fig. 3.

The complete absence of long gaps along the vertical axis in all diagrams indicates the successful implementation of the attack. No device was able to enter the sleep mode with the exception of HHL mode between 8 and 12 s. All packets sent by the attacking parasitic device were recognized as normal traffic and forwarded to victim hosts. Generally, except for the subtleties of the implementation and the goal pursued, this attack is similar to conventional DoS attacks on wireless sensor networks. However, the attack is easy to detect, since during the transmission of normal traffic, there is an overlap of attacking traffic, which can be seen in the transcript, and, since the interval of attacking traffic is fixed, it is visually possible to filter out attacking traffic, even if only the time of receipt of the message is available. In the diagram, the attacking traffic can be observed among the visible traffic in the

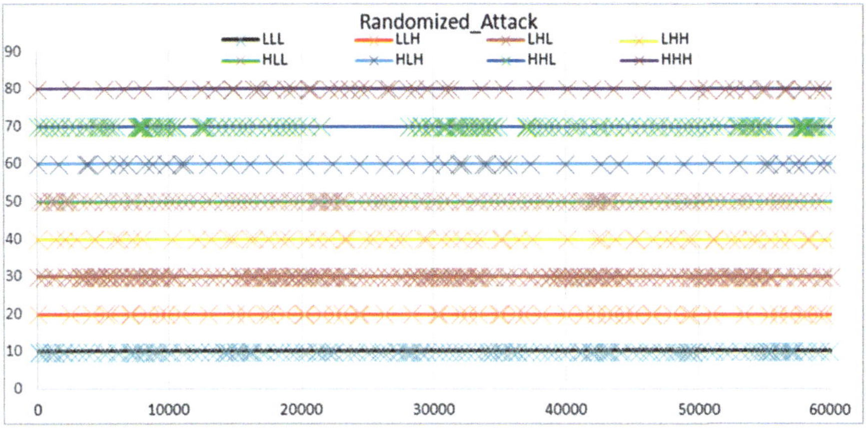

Fig. 5 Attack traffic visualization—randomized attack

HLH mode. Moreover, when applied with a different scaling in the diagram, this will be visible in other modes as well.

(2) Denial-of-Sleep attack variation 2 (randomized attack)

It is assumed that in order to increase the secrecy of the attack, the attacker introduces certain randomness in determining the interval between messages, which is the same as in normal traffic (Fig. 5).

Based on the specificities of Fig. 5, it can be observed that in the intervals between bunches, attacking traffic behaves much more chaotically. It is now possible to determine visually only the very fact of the attack itself (and not the time of its activity) since during the data transfer, there is still an overlap of normal and attacking traffic. However, it becomes complicated to distinguish one traffic from the other one.

(3) Denial-of-Sleep attack—variation 3 (enhanced attack)

This type of attack includes the interception and analysis of traffic from legitimate users and the direct filling with attacking messages of those periods of time in which there is no transmission of legitimate messages. Therefore, if the intruder sends attack messages only after the end of the next bunch of legitimate messages, then the victim node will assume that it receives a longer bunch and will not be able to distinguish the attacker's messages from the normal traffic. In addition, on the attacked node, it will be difficult to detect the very fact of the attack, since the delay between the normal bunch and the attacker's bunch can be taken as a normal message delay caused by the peculiarity of the data transmission medium (Fig. 6).

Accordingly, as soon as the legitimate node stops transmitting messages, the attacker starts transmitting data with the same message interval in a bunch and with the same time deviations. In addition, during the attack, the attacker also takes into

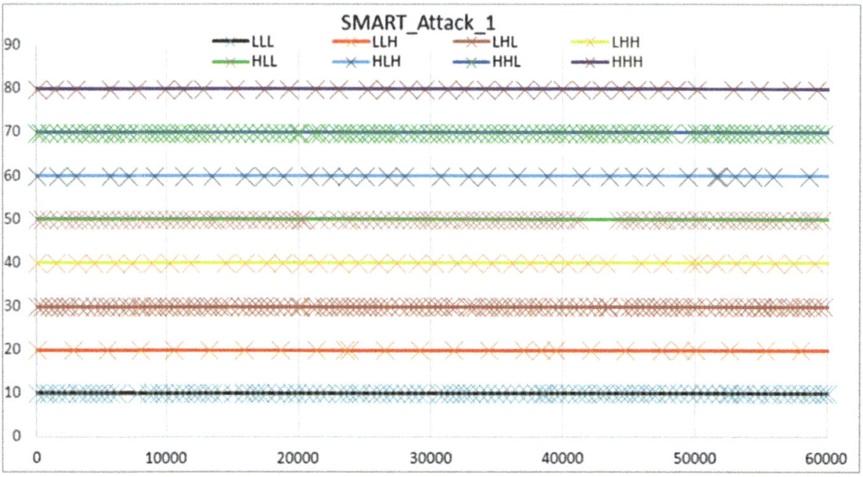

Fig. 6 Attack traffic visualization enhanced attack (variation 3)

account a minimum possible delay of messages between bunches in order to ensure that the sleep mode of the attacked node has minimum allowable length. This leads to a decrease in the battery life of the attacked node, but not its exhaustion.

Figure 6 shows that, in comparison with the previously considered attack types, there are no visible overlaps along the horizontal axes. Therefore, it is difficult to identify the attack by visual means. At the same time, the intervals between the bunches become much smaller than in the absence of an attack. Thus, in comparison with the previous types of attacks, this attack is less effective but much more stealthy.

(4) Denial-of-Sleep attack—variation 4 (enhanced attack)

The type of attack has been modeled where the attacker's messages arrive less frequently but with an interval within limits for normal traffic. To simulate such an attack, the attacker's messages are stretched to the maximum allowable interval between messages in a bunch. In Fig. 7, the diagram shows sections of shrinkage and expansion, indicating the transition between the normal traffic and the attacking one. However, we should note that the number of messages decreases which makes it difficult to visually detect the beginning of an attack by the number of the maximum allowable number of messages in the bunch. Moreover, due to the increased interval between messages in a bunch, it is now more difficult to determine the beginning of the next bunch, since visually the interval between bunches is no longer so noticeable.

(5) Denial-of-Sleep attack—variation 5 (enhanced attack)

And finally, let us consider the last type of Denial-of-Sleep attack. The main feature of such attack is that it does not require sending new messages to the

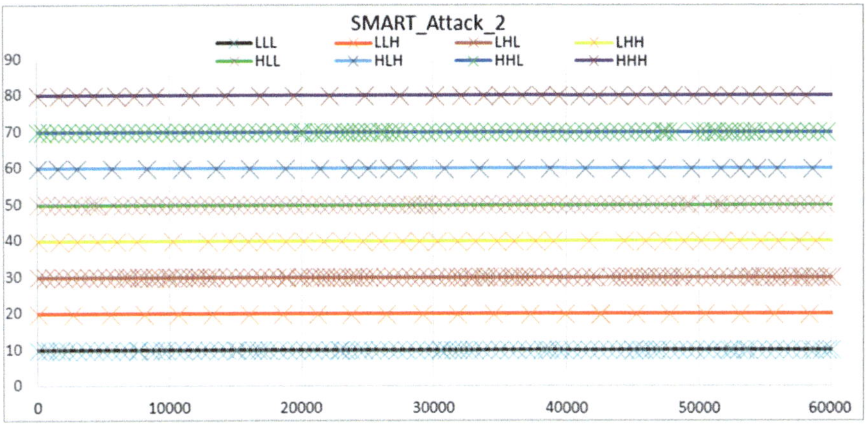

Fig. 7 Attack traffic visualization enhanced attack (variation 4)

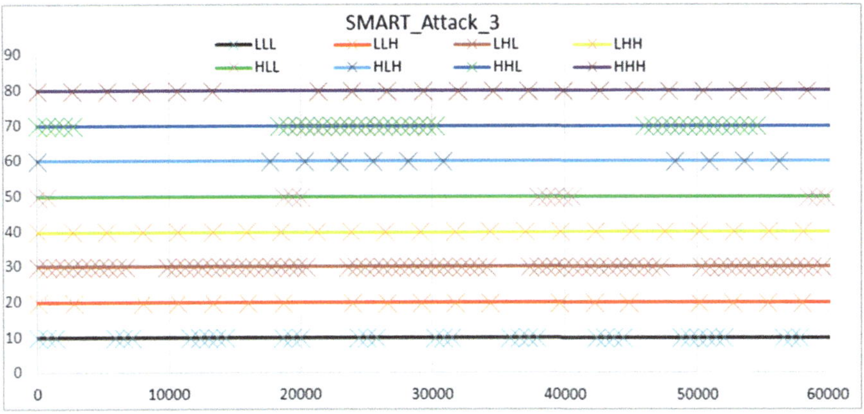

Fig. 8 Attack traffic visualization enhanced attack (variation 5)

network. It is assumed that the attacker is able to intercept and delay normal traffic messages for some time, while keeping a larger but acceptable interval between messages in a bunch. Therefore, resulting from such attack, the interval between bunches decreases, and consequently the time spent by the victim in the sleep mode decreases.

It is to be noted that LHH-type traffic was completely deprived of intervals. The intervals in HHH and LHH were shortened significantly. Thus, using the same number of messages as in unmodified normal traffic, within the allowable intervals of messages in the bunch, the attacker was able to simulate a Denial-of-Sleep attack quite effectively (Fig. 8).

5 Development of the Detection Mechanism and Experiments

Figure 9 shows a general scheme for studying the processes of detecting Denial-of-Sleep attacks in WSN. The scheme consists of the following seven main phases:

(1) Analysis of available logs, including records of the normal functioning of an autonomous network node and a node under various variations of a Denial-of-Sleep attack.
(2) Formation of a feature space and extraction of specific values of attack features. Selection of training and test samples.
(3) Formation of several datasets and check of their balance. A two-class classification is used to check the presence or absence of Denial-of-Sleep attacks, using all the initial data. Besides, these methods are used to make a classification of each of the considered types of attackers differing in the degree of the attack effectiveness and its secrecy. Thus, six datasets are formed, where norm denotes the logs of the normal functioning of the device, while atti is the logs of the device that is under the influence of the attacking type i:

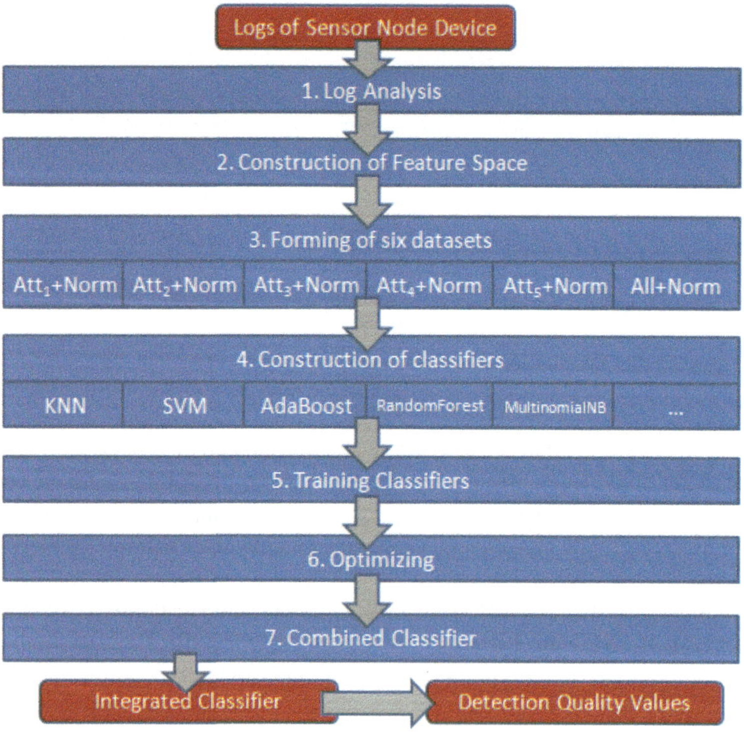

Fig. 9 Stages of the study of Denial-of-Sleep attack detection processes

$$\{\bigcup_{i=1}^{5}(norm, att_i)\} \cup \{(norm, \sum_{i=1}^{5} att_i)\}$$

(4) Formation of classifiers based on a number of machine learning methods, namely, KNeighborsClassifier, SVM, LogisticRegression, AdaBoost, and GaussianNB [1].
(5) Training of classifiers on the training samples of six available datasets. The training method hyperparameter values are set to default ones. Testing of the classifiers and calculating quality indicators for detecting attacks.
(6) Selection of the best classification methods for each dataset by comparing the values of the detection quality indicators.
(7) Integral classification and evaluation on the total amount of data, with account taken of the attacks of all five variations.

We have analyzed the available logs including records of normal functioning of an autonomous network node and a node under five different variations of the Denial-of-Sleep attack. The duration of logging for each of the attack and normal operation logs is approximately 10 min. There are different numbers of messages in each of the logs. For the logs of normal functioning, the number of messages varies from 115 to 826 pieces, depending on specific values of 3 assigned modeling parameters. For logs describing mixed traffic, this value varies from 96 to 1980 messages. The timestamps for each entry were normalized with respect to zero time point.

To detect attacks by using intelligent methods, a feature space consisting of $N = 50$ features was specified. Each of the samples is a chain of timestamps, presenting the time of arrival of the next 50 messages that sequentially arrived at the victim node. The division into training and testing samples was made in the ratio of 0.75 to 0.25, respectively. Labels of two classes are used, namely, 0 for normal traffic and 1 for traffic containing attacking messages.

A fragment of a dataset built for mixed traffic for a simple Denial-of-Sleep attack is shown in Fig. 10. For universality and usability, the names of the features indicated in the top line are conventionally denoted by nonnegative integers. In particular, Fig. 10 shows that the first eight samples were obtained by simply splitting the input sequence of timestamps into N vectors. One can also see that, starting at sample number 8, the next sequence of timestamps follows.

To develop intelligent methods for detecting Denial-of-Sleep attacks, software implementations of machine learning the following methods from the Scikit-learn library were used: KNeighborsClassifier, SVC, LogisticRegression, AdaBoost, and GaussianNB. They were chosen due to the fact that they have been proven as a way to solve classification problems in various typical application fields. At the same time, their diversity and different principles they are built on allow us to expect finding among them some quite promising ones that are able to give acceptable values of classification quality indicators. For each of the five types of Denial-of-Sleep attacks, these machine learning methods were run on the training

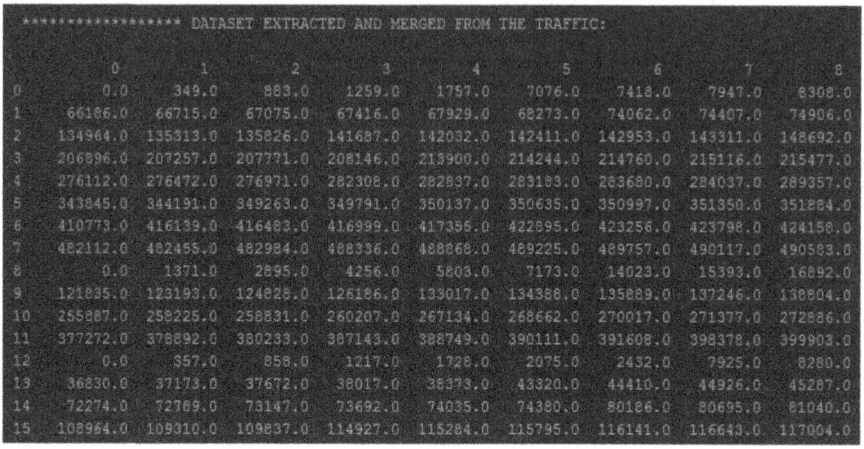

******************** DATASET EXTRACTED AND MERGED FROM THE TRAFFIC:

	0	1	2	3	4	5	6	7	8
0	0.0	349.0	883.0	1259.0	1757.0	7076.0	7418.0	7947.0	8308.0
1	66186.0	66715.0	67075.0	67416.0	67929.0	68273.0	74062.0	74407.0	74906.0
2	134964.0	135313.0	135826.0	141687.0	142032.0	142411.0	142953.0	143311.0	148692.0
3	206896.0	207257.0	207771.0	208146.0	213900.0	214244.0	214760.0	215116.0	215477.0
4	276112.0	276472.0	276971.0	282308.0	282837.0	283183.0	283680.0	284037.0	289357.0
5	343845.0	344191.0	349263.0	349791.0	350137.0	350635.0	350997.0	351350.0	351884.0
6	410773.0	416139.0	416483.0	416999.0	417355.0	422895.0	423256.0	423798.0	424158.0
7	482112.0	482455.0	482984.0	488336.0	488868.0	489225.0	489757.0	490117.0	490583.0
8	0.0	1371.0	2895.0	4256.0	5803.0	7173.0	14023.0	15393.0	16892.0
9	121835.0	123193.0	124828.0	126186.0	133017.0	134388.0	135889.0	137246.0	138804.0
10	255887.0	258225.0	258831.0	260207.0	267134.0	268662.0	270017.0	271377.0	272886.0
11	377272.0	378892.0	380233.0	387143.0	388749.0	390111.0	391608.0	398378.0	399903.0
12	0.0	357.0	858.0	1217.0	1728.0	2075.0	2432.0	7925.0	8280.0
13	36830.0	37173.0	37672.0	38017.0	38373.0	43320.0	44410.0	44926.0	45287.0
14	72274.0	72789.0	73147.0	73692.0	74035.0	74380.0	80186.0	80695.0	81040.0
15	108964.0	109310.0	109837.0	114927.0	115284.0	115795.0	116141.0	116643.0	117004.0

Fig. 10 A fragment of a dataset built for mixed traffic for a simple Denial-of-Sleep attack

Table 1 Detection quality indicators for Simple Denial-of-Sleep attack

ML method	Accuracy	Precision	Recall	f1-measure
KNeighborsClassifier	0.96	[0.89, 0.98]	[0.89, 0.98]	[0.89,0.98]
SVC	0.89	[1.0,0.88]	[0.33,1.0]	[0.5,0.93]
LogisticRegression	0.84	[0.5, 0.91]	[0.56, 0.89]	[0.53, 0.9]
AdaBoostClassifier	**0.98**	**[1.0, 0.98]**	**[0.89, 1.0]**	**[0.94, 0.99]**
GaussianNB	0.85	[0.56,0.91]	[0.56,0.91]	[0.56,0.91]

Table 2 Detection quality indicators for Randomized Denial-of-Sleep attack

ML method	Accuracy	Precision	Recall	f1-measure
KNeighborsClassifier	0.92	[0.78, 0.96]	[0.78, 0.96]	[0.78, 0.96]
SVC	0.9	[1.0, 0.9]	[0.44,1.0]	[0.62,0.95]
LogisticRegression	0.78	[0.38,0.9]	[0.56,0.82]	[0.45,0.86]
AdaBoostClassifier	**1.0**	**[1.0, 1.0]**	**[1.0, 1.0]**	**[1.0, 1.0]**
GaussianNB	0.85	[0.56,0.91]	[0.56,0.91]	[0.56,0.91]

sets. When checking the constructed classifiers on test samples, the indicators of accuracy, precision, recall, and f1-measure were calculated. Tables 1, 2, 3, 4, and 5 show the values of these indicators for each type of attack for various machine learning methods. For precision, recall, and f1-measures, the values are presented independently for each class, i.e., a value for the norm and a value for the attack. The indicator values are given in a format of up to two digits after the decimal point. As part of the experiments, the optimization included finding out, enumerating, and comparing on independent test data the best values of detection quality indicators for each type of attack. In Tables 1–6 the best values of the used indicators are highlighted in bold.

Table 3 Detection quality indicators for Smart Denial-of-Sleep attack (variation 3)

ML method	Accuracy	Precision	Recall	f1-measure
KNeighborsClassifier	0.77	[0.73,0.78]	[0.53,0.89]	[0.62,0.83]
SVC	0.65	[0.50,0.76]	[0.60,0.68]	[0.55, 0.72]
LogisticRegression	0.79	[0.88,0.77]	[0.47,0.96]	[0.61,0.86]
AdaBoostClassifier	**0.93**	**[0.93, 0.93]**	**[0.87, 0.96]**	**[0.90, 0.95]**
GaussianNB	0.63	[0.47,0.73]	[0.53,0.68]	[0.5,0.7]

Table 4 Detection quality indicators for Smart Denial-of-Sleep (variation 4)

ML method	Accuracy	Precision	Recall	f1-measure
KNeighborsClassifier	0.87	[1.0,0.84]	[0.58,1.00]	[0.74,0.92]
SVC	0.77	[0.80,0.76]	[0.33,0.96]	[0.47,0.85]
LogisticRegression	0.79	[0.83,0.79]	[0.42,0.96]	[0.56,0.87]
AdaBoostClassifier	**0.97**	**[0.92, 1.0]**	**[1.00, 0.96]**	**[0.96, 0.98]**
GaussianNB	0.79	[0.75,0.81]	[0.5,0.93]	[0.6,0.86]

Table 5 Detection quality indicators for Smart Denial-of-Sleep attack (variation 5)

ML method	Accuracy	Precision	Recall	f1-measure
KNeighborsClassifier	0.74	[0.79,0.67]	[0.79,0.67]	[0.79,0.67]
SVC	0.61	[0.62,0.5]	[0.93,0.11]	[0.74, 0.18]
LogisticRegression	0.39	[0.5,0.31]	[0.36, 0.44]	[0.42, 0.36]
AdaBoostClassifier	**1.0**	**[1.0, 1.0]**	**[1.0, 1.0]**	**[1.0, 1.0]**
GaussianNB	0.61	[0.63,0.50]	[0.86, 0.22]	[0.73, 0.31]

Table 6 Detection quality indicators for Denial-of-Sleep attack (integrated classifier)

ML method	Accuracy	Precision	Recall	f1-measure
KNeighborsClassifier	0.91	[0.2,0.93]	[0.08,0.97]	[0.12,0.95]
SVC	0.91	[0.0,0.93]	[0.0,0.99]	[0.0,0.96]
LogisticRegression	0.92	[0.4,0.94]	[0.17,0.98]	[0.24,0.96]
AdaBoostClassifier	**0.97**	**[0.89, 0.97]**	**[0.67, 0.99]**	**[0.76, 0.98]**
GaussianNB	0.87	[0.31,0.97]	[0.67,0.88]	[0.42,0.92]

In addition, a general integral classification was performed on a combined dataset. The results of the final classification are shown in Table 6.

For all five types of attacks and their combination among the used training methods, the AdaBoostClassifier method was determined as the best. Table 7 shows the final weighted average values of the detection quality in the form of f1-measure. This indicator represents an averaged over all the considered classes and should be considered as a target one.

Table 7 Weighted avg f1-measure for AdaBoostClassifier

	Weighted avg f1-measure
Simple Denial-of-Sleep attack	0.98
Randomized Denial-of-Sleep attack	1.00
Smart Denial-of-Sleep attack (variance 3)	0.98
Smart Denial-of-Sleep attack (variance 4)	0.95
Smart Denial-of-Sleep attack (variance 5)	1.00
Integrated classification	0.97

6 Discussion

As an analysis of the results obtained experimentally, we can note that the final values of the detection quality seem to be quite high. In fact, for all five variations of Denial-of-Sleep attacks, the integral weighted average values of the f1-measure exceed 0.95. At the same time, the data obtained in their current form does not allow making unambiguous correlation between the quality of attack detection and the degree of its secrecy. In accordance with the data illustrated in Figs. 3, 4, 5, 6, 7, and 8, the simple Denial-of-Sleep attack and the randomized one are the most effective in terms of battery drain acceleration. All three cases of a smart attack show that an attacker's attempt to hide the fact of the attack from detection tools leads to a decrease in its effect.

Despite rather high values of the calculated indicators, it makes sense to use the following as possible ways to further improve the quality of Denial-of-Sleep attack detection. As basic ways, it will be profitable to expand the list of classification methods as well as to apply the automatic enumeration of hyperparameters and testing the quality by using GridSearch mechanism.

In addition, it is possible to reproduce the learning process on extended sets of initial data. In particular, sampling can be used to increase the balance of data between classes, which is important especially for some machine learning methods. It can be based on a partial shift of the feature vector closer to the beginning of the vector by k pieces, filling the remaining k features at the end of the vector with values from the beginning of the next time sample. In this case, with such sampling, the very first and latest vectors will be discarded. In conditions of transformations of time sequences of events to the group of features used, it is necessary to control the absence of overlaps of the same data fragments that can fall into both training and test samples. Checking and avoiding such intersections will make it possible not to get overly optimistic classifier estimates due to insufficient independence of the tested data from the training data. In particular, this will be aimed at avoiding the effect of overfitting. Preliminary experimental data on the use of such sampling to detect a Denial-of-Sleep by attack using AdaBoostClassifier allowed us to slightly increase the values of detection quality indicators. Other sampling techniques are also possible, allowing us to more accurately select the desired samples, including samples belonging to certain classes. In case of insufficiency of the available volumes of initial data, it is also reasonable to use cross validation along with sampling.

The possibilities for increasing the number of samples in the initial data, which still needs to be tested in practice, include a decrease in the value of the parameter N, which is responsible for cutting the logs into samples. It is to be noted that the total amount of initial data will not increase as a result and there is no guarantee that the detection quality will increase or even not decrease. When applying certain methods, such as the nearest neighbor method or logistic regression, it is also possible to improve the detection quality due to additional transformations and normalization of features, for example, by transformation of data to a certain single scale by using mechanisms of MinMaxScaler, StandardScaler, etc.

In addition, an increase in the detection quality can also be achieved by combining classifiers through using voting and stacking techniques. Such techniques can be used both for detecting specific types of Denial-of-Sleep attacks and for integral classification.

The Denial-of-Sleep attack detection mechanisms built in this work were trained and tested on eight intensity modes of each of the analyzed five types of attacks. Despite the fact that during the experiments, the principle of independence of data from test samples and data from training samples was strictly followed; samples of each of the modes fell into both training and test samples. This circumstance allows us to establish the fact that in practice the constructed classifiers will be guaranteed to work with the preset quality only if they are used to detect a Denial-of-Sleep attack with one of the eight operation modes we have experienced (or in a mode close to one of the established ones). In modes other than those indicated, it is most likely the classifiers will also give quite appropriate results, but the quality of detection may decrease to some extent.

Therefore, a possible improvement in the learning process will be a modification of the partitions of the initial data into training and test samples. In particular, it would be expedient to allocate one or several attack simulation models for general testing. And during the training process, the samples that make up the attack data of these test modes should not be included into the training samples. In other words, it is proposed not only to identify the samples of the training and testing sets from each other but also to isolate the modes. Specifically, some of the modes will go entirely into the training set and the rest into the testing set. At first, this modification can lead to a decrease in the quality of detection because the testing and training data will have structures that differ from each other to a greater extent. Nevertheless, the use of the methods described above to improve the quality of detection will help to prevent this effect.

In general, we can note that the experiments were based on log records extracted from the traffic of WSN devices that communicate using the ZigBee protocol with a number of technological limitations and specificities. In particular, these include a rather low range of wireless signal, use of dynamic routing based on static addressing of nodes, and absence of any protection tools in version 2 of the protocol. Despite all this, modeling of Denial-of-Sleep attacks and the possibility of generalizing the obtained results to other WSN communication protocols, such as LoRaWAN, LPWAN, Sigfox, etc., seem to be quite reasonable. This is true because the basic principles of organizing such Denial-of-Sleep attacks remain quite universal and feasible in relation to a wide range of IoT devices that operate from autonomous power supplies and support dynamic power efficiency mode change [5].

The peculiarities of the modeling performed in this work include its hybrid nature. It lies in the combination of full-scale modeling of the WSN and the processes of its functioning on the available hardware/software equipment, on the one side, and simulation modeling to form vectors of attacks with different frequency characteristics, on the other.

7 Conclusions

In this work, modeling of Denial-of-Sleep attacks on wireless sensor network nodes operating from autonomous exhaustible energy sources was performed. The modeling was fulfilled within the framework of the applied scenario of the operation of a ZigBee mesh topology network with Digi XBee version 2 nodes, Arduino microcontrollers, and other electronic components. This testbed modeled a Denial-of-Sleep spoofing-based attack on a network node. Affected by attacking messages arriving at a certain flow rate, the node cannot switch to a power-saving mode. As a result, the node's battery is depleted at a much faster rate than the node's normal operation. The attacking traffic was recorded, and this data was used as the initial data for building the detection mechanism. Denial-of-Sleep attacks are detected applying machine learning methods using a series of features with normalized values. Precision, recall, and f1-measure, as well as the indicator of the weighted average value of the f1-measure were chosen as the main indicators of the detection quality for the selection of specific training models. The experiments were carried out on normal traffic and five types of attacking traffic, which differ in the ease of implementation of the attack, its effect, and secrecy. Among selected training models, the best values of indicators of the detection quality of all the considered variations of Denial-of-Sleep attacks were obtained using AdaBoostClassifier model. On average, the value of f1-measure was about 0.97, which confirms the validity of using such a model to effectively detect this type of attack.

As a part of further work in this field, it is planned to study alternative machine learning methods and evaluate them. In addition, it is prescheduled to expand the structure and volume of the initial data, as well as to expand the list of types of Denial-of-Sleep attacks with the possibility of transferring the results of experiments to other types of IoT systems.

References

1. Scikit-learn library. machine learning in python. https://scikit-learn.org. Accessed: 2022-03-14.
2. Tapalina Bhattasali and Rituparna Chaki. AMC model for denial of sleep attack detection. 03 2012.
3. Angelo T Capossele, Valerio Cervo, Chiara Petrioli, and Dora Spenza. Counteracting denial-of-sleep attacks in wake-up-radio-based sensing systems. In *2016 13th Annual IEEE International Conference on Sensing, Communication, and Networking (SECON)*, pages 1–9. IEEE, 2016.
4. Flvia Delicato, Salmo H., Claudio De Farias, Luci Pirmez, Silvana Rossetto, Paulo Rodrigues, and Luiz Carmo. Intrusion detection system for wireless sensor networks using danger theory immune-inspired techniques. *International Journal of Wireless Information Networks*, 19:179, 03 2012.
5. Vasily Desnitsky and Igor Kotenko. Simulation and assessment of battery depletion attacks on unmanned aerial vehicles for crisis management infrastructures. *Simulation Modelling Practice and Theory*, 107(5):1263–1278, 2021.
6. Vasily Desnitsky, Igor Kotenko, and Danil Zakoldaev. Evaluation of resource exhaustion attacks against wireless mobile devices. *Electronics*, 8(5), 2019.
7. Vasily A. Desnitsky, Igor V. Kotenko, and Nikolay N. Rudavin. Protection mechanisms against energy depletion attacks in cyber-physical systems. In *2019 IEEE Conference of Russian Young Researchers in Electrical and Electronic Engineering (EIConRus)*, pages 214–219, 2019.
8. G. S. Dhunna and I. Al-Anbagi. A low power wsns attack detection and isolation mechanism for critical smart grid applications. *Sensors Journal*, 19(13):5315–5324, 2019.
9. Noor Mohd, Annapurna Singh, and H. Bhadauria. A novel svm based ids for distributed denial of sleep strike in wireless sensor networks. *Wireless Personal Communications*, 111, 04 2020.
10. N. Nishanth and A. Mujeeb. Modeling and detection of flooding-based denial-of-service attack in wireless *ad hoc* network using bayesian inference. *IEEE Systems Journal*, 15(1):17–26, 2021.
11. Abdul Razaque, M. Abdulghafour, and Meer Jaro Khan. Detection of selfish attack over wireless body area networks. In *2017 IEEE Conference on Open Systems (ICOS)*, pages 48–52, 2017.
12. Ekereuke Udoh and Vladimir Getov. Performance analysis of denial-of-sleep attack-prone mac protocols in wireless sensor networks. In *2018 UKSim-AMSS 20th International Conference on Computer Modelling and Simulation (UKSim)*, pages 151–156, 2018.
13. Jason Uher, Ryan Mennecke, and Sam Farroha. Denial of sleep attacks in bluetooth low energy wireless sensor networks. pages 1231–1236, 11 2016.

Machine Learning Methodologies for Preventing Malware Obfuscation

Vincenzo Carletti, Alessia Saggese, Pasquale Foggia, Antonio Greco, and Mario Vento

1 Introduction

Malware is among the most relevant cyber threats; thousands of malware are constantly delivered with the purpose of infecting as many devices as possible by exploiting the large variety of systems and software platforms that pervade our everyday life. Indeed, the widespread diffusion of IoT devices has considerably increased the attack surface of malware; devices that have been traditionally targeted since the 1970s [2], like servers and personal computers, have become less interesting for malware authors than smartphones and smart watches but also smart thermostats or light bulbs [14, 50]. Recent reports of cybersecurity companies highlight how much this threat is growing over the years [12, 20, 52].

The variety and the large availability of targets are not the only relevant aspects of the problem. Malware is mutable, adaptable, and suitable to be easily camouflaged, making the detection even more challenging. Therefore, cybercriminals can exploit a large arsenal of techniques aimed at hiding malicious software to antiviruses. As a consequence, malware detection and classification systems are demanded to constantly evolve to deal with such a mutable scenario, pushing researches to focus on new approaches and techniques. Several methods have been proposed in the literature over the last decades, and they can be categorized in three main families: (i) static approaches that analyze executable or source files without running the programs; (ii) dynamic approaches that examine the behavior of a software by executing it in safe environments, like sandboxes; and (iii) hybrid approaches combining both static and dynamic analysis to obtain a more reliable outcome.

V. Carletti (✉) · A. Saggese · P. Foggia · A. Greco · M. Vento
Department of Computer Engineering, Electrical Engineering and Applied Mathematics, University of Salerno, Salerno, Italy
e-mail: vcarletti@unisa.it; asaggese@unisa.it; pfoggia@unisa.it; agreco@unisa.it; mvento@unisa.it

L. Fotia et al. (eds.), *Security, Trust and Privacy Models, and Architectures in IoT Environments*, Internet of Things, https://doi.org/10.1007/978-3-031-21940-5_6

A detailed review of the most relevant approaches belonging to each family has been proposed by Aslan et al. in [4].

Many antivirus products still heavily rely on *signature-based* methods; a set of static features, such as byte sequences, assembly instructions, n-grams, lists of system calls, and so on, are extracted from an executable file to compose a signature describing the executable; the signature is then compared against those are contained in a large dataset of malware signatures to find a match. Since this strategy often fails to detect malware that was never seen before (and thus is not in the database), the most advanced antivirus systems usually extend the basic signature-based analysis with the addition of heuristics or behavior-based approaches to make the detection more reliable [4].

Machine learning methods have been adopted to perform both static and dynamic analysis; most of them are based on hand-crafted features, to which they apply traditional learning algorithms such as k-nearest neighbor (k-NN), Bayesian networks (BN), random forests (RF) [45, 55, 56], or support vector machines (SVM) [7, 31]. Very recently also deep learning methods have been proposed [38, 39, 60]. Among them, a popular trend is based on the use of convolutional neural networks (CNNs) to perform the static analysis [19, 25, 32, 49, 53, 58]; these methods propose to transform a binary executable file into an image and then reuse state-of-the-art CNNs to process the image and learn the features to distinguish among different families of malware (Fig. 1). For the sake of clarity, the idea of using an image-based representation of malware is not so recent; indeed, it was firstly proposed by Nataraj et al. [36], but the classification was performed using traditional methods: GIST features were extracted from the images, and then a k-NN was used to classify the images. The renewed interest in using such a representation is encouraged by the impressive performance achieved by CNNs on several image analysis tasks and motivated by the availability of ready to use state-of-the-art networks that can be directly applied to malware classification through transfer learning, such as ResNet50 [22], InceptionV3 [51], VGG16 [48] and MobileNet [46]. Beyond the convenience in adopting CNNs for malware classification, a relevant question, to be answered before considering these methods really suitable to substitute or support traditional approaches, is how robust are they against malware obfuscation. Indeed, it is has been demonstrated that machine learning methods, as well as deep neural networks in particular, can be easily fooled by properly generated corrupted samples [5]. Some papers have faced the problem by applying carefully crafted noise on the images [37, 60], but without demonstrating or discussing if the corrupted samples still behave as the original ones, or even if they are still executable at all. However, in the real world, a cybercriminal intending to evade an antivirus would apply some form of obfuscation directly on the executable in a way that would obviously not prevent its execution; thus a better approach to evaluate the robustness of a detection method is to corrupt the executable using the techniques proposed in [8, 15, 29]. These papers propose sophisticated obfuscation techniques based on optimization or on junk code insertion, but they do not consider simpler forms of obfuscation, more easily implemented by an attacker, that may still prove to be effective against a detection method.

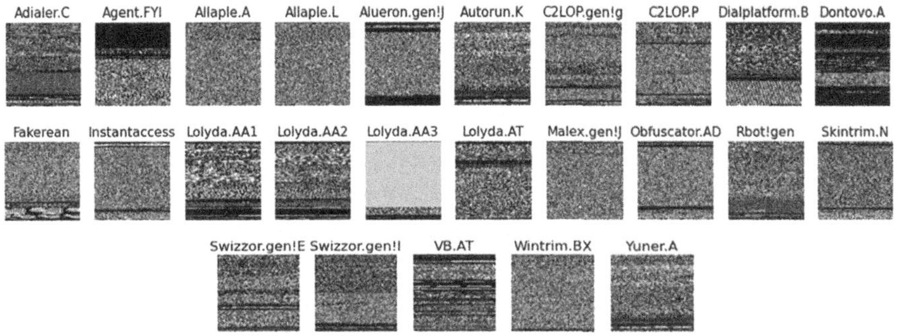

Fig. 1 Example of images extracted from malware binary that are contained in the dataset MalImg [36]

In this paper we analyze how some simple obfuscation techniques, which will be discussed in more details in Sect. 3.2, may affect the accuracy of state-of-the-art methods for malware classification, including both methods using hand-crafted features and traditional classifiers, i.e., LightGBM [26] and XGBoost [9], and image-based methods using CNNs like ResNet50 [22], XceptionV3 [11, 51], VGG16 [48], and MobileNetV2 [46]. The robustness analysis has been conducted using a recently published dataset composed of 10 million disarmed malware samples, namely, SOREL-20M. In order to analyze if the methods are still able to distinguish malicious from benign samples when the obfuscation is applied, we have extended the datasets by adding benign samples collected from public repositories of freeware. The analysis has demonstrated that most of the considered methods are very sensitive with respect to some basic obfuscation techniques and are not able to discriminate between an obfuscated malware and a benign software, making them completely useless as malware detection systems. Nevertheless, some of them have shown a good resilience against the obfuscation with an acceptable drop of accuracy when distinguishing between malware and benign samples but losing the capability to properly discriminate among different families of malware.

2 Machine Learning Methods for Malware Classification

As previously introduced, the analysis we propose considers both traditional machine learning methods that are still widely adopted by the research community, such as LightGBM [26] and XGBoost [9], and the most recent image-based approaches that use CNNs, like ResNet50 [22], XceptionV3 [11, 51], VGG16 [48], and MobileNetV2 [46]. In this section we are going to discuss in more details the main differences among these methods.

2.1 Traditional Machine Learning Methods

Explainability is still a mandatory requirement for several malware detection systems since cybersecurity experts need to be aware of the decision taken by the algorithm. For this reason, random forests [23], supervised learning methods based on an ensemble of multiple decision trees, are among the most adopted machine learning methods to classify malware. Some recent papers [1, 18, 34, 35, 40, 47] discuss the high accuracy achieved by random forest methods in detecting and classifying malware, both on Windows PE and on Android executables. In particular, what is highlighted in [18] and also supported by other works [3, 43, 47, 56] is the effectiveness of using gradient boosting [16, 17], a machine learning technique where the decision trees are arranged as a sequence, so that each classifier tries to reduce the error of its predecessors, with the aim of minimizing a loss function. Among these approaches those are mostly used and suggested to the task at hand are LightGBM [26] and XGBoost (Extreme Gradient Boosting) [10], two gradient-boosting frameworks widely adopted, thanks to capability to achieve very high accuracy, efficiency, and scalability.

We consider random forests as traditional machine learning approaches because they use hand-crafted features properly selected by a domain expert. To this purpose, in the literature there are many features that have been proposed to characterize executable files. In this paper, we use a very large set of features proposed in [3] to build the EMBER (Endgame Malware BEnchmark for Research) dataset; the procedure to extract the features has been well described in the paper and uses the open source library LIEF (Library to Instrument Executable Formats). The procedure assumes that the input binary is a Windows Portable Executable (PE) and produces a representation composed of eight groups of raw features including values parsed from the PE headers like timestamp, architecture, imports, resources, sections and so on, format-agnostic histograms and statistics about strings (average length, count, entropy, path counts, etc.).

All these raw features are purposely human understandable, and the authors also provide a procedure to extract numeric features from the raw features in order to get the vectors needed for training machine learning models. Besides the features, the EMBER dataset also includes a unique identifier for the each file computed using SHA256, time information about when the file has been analyzed for the first time, and a label that states if the file is a malware or not.

2.2 Image-Based Malware Detection

Differently from the random forest, CNNs are able to learn the best vectorized representation of an input file during the training process, without any human aided feature selection. Since they are designed to process images, the only preprocessing phase required is the conversion of binary files into images.

Table 1 The table reports the width of the image with respect to the size of the executable. We report the file size range, on the left, and its corresponding image width on the right

File size range	Image width
<10kB	32
10kB–30kB	64
30kB–60kB	128
60kB–100kB	256
100kB–200kB	384
200kB–500kB	512
500kB–1000kB	768
>1000kB	1024

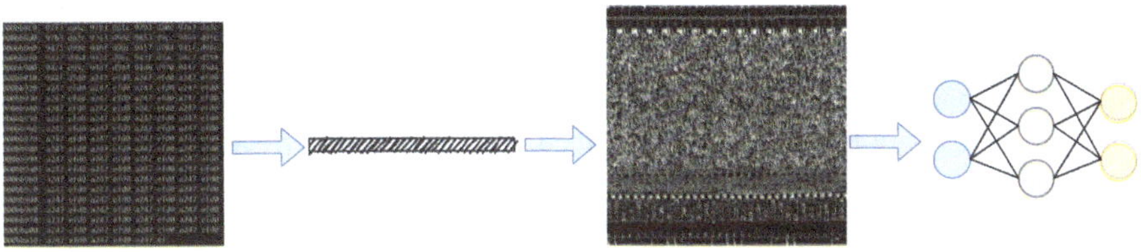

Fig. 2 Overview of the general workflow of an image-based malware classification approach. The binary file is represented as an array of integers (values between 0 and 255), which is then arranged in a rectangular gray-scale image, whose width depends on the original size of the malware. Finally, the image is fed to a CNN trained for malware classification

The conversion process has been firstly proposed in [36] and adopted by different image-based approaches [13, 30, 36, 54, 57]. In particular, the binary executable file (or sometimes a subset of it) is interpreted as an array of integer values ranging from 0 to 255. Then the sequence of integers is reshaped into a two-dimensional matrix that represents the grey scale image. Moreover, in [36] the authors have experimentally evaluated and proposed how to manage the width of the matrix; they propose to vary this parameter, depending on the whole image size or the size of the file as shown in Table 1. Finally, the number of pixels composing the height is trivially computed by dividing the file size by the width. The overall conversion process is briefly shown in Fig. 2.

Once the image is obtained, a CNN can process it and discriminate between malware and benign samples or distinguish among different family of malware. In order to make the CNN able to perform these tasks, it is possible to use networks pre-trained to recognize objects and specialize them on malware images through transfer learning. To this purpose ResNet [22] is one of mostly used CNNs for malware detection and classification [6, 27, 28, 41, 59]. Together with ResNet other state-of-the-art CNNs have been used; in particular, in [42] it is demonstrated that VGG16 can achieve a very high accuracy in distinguishing 20 different families of malware represented as images, while in [33] the authors have compared VGG16 and Xception [11, 51] concluding that the latter outperforms VGG16 on malware image classification. More recently in [8], the authors have compared the accuracy and the robustness of ResNet-50, XceptionV3, VGG16, and MobileNetV2 [46] suggesting to use this latter due to its robustness against the obfuscation through junk code insertion.

For the sake of completeness, we provide some few details about the architectural differences of the considered CNNs. VGG16 [48] is the most simple and widely used CNN for image classification tasks prosed by the Visual Geometry Group of University of Oxford; it has a traditional architecture with a fixed input size of 224×224 pixels that is passed through a sequence of convolutional layers (usually five layers) followed by three fully connected layers and a softmax. ResNet is one of the most powerful deep neural networks; the most relevant characteristic of its architecture is the use of residual blocks: each convolutional layer of the network is connected not only with its successor but also with k-steps ahead through skip connections. The advantage in using residual blocks is that they allow to effectively train a deeper architecture reducing the effect of the vanishing gradients during the backpropagation. There exist many variants of the ResNet [22] architecture that differs mainly on the number of layers; in particular ResNet50 has 50 layers. MobileNetV2 [46] is the evolution of the MobileNet [24], a CNN designed to be efficient on mobile devices by reducing the model size and the complexity, thanks to the introduction of the *depthwise separable convolution*: a convolution divided per color channel (depthwise convolution), followed by a pointwise convolution used to reduce the dimensionality. MobileNetV2 [46] extends the original architecture by adding two new concepts: inverted residual blocks and linear bottlenecks. In details, the overall architecture of MobileNetV2 is composed of a fully convolution layer with 32 filters, followed by 19 *bottleneck layers* back connected by residual connections. Each block begins with a narrow layer that is expanded and then squeezed again at the end of the block. Here lies the main difference between the original residual blocks used in ResNet and the inverted residual blocks of MobileNetV2, indeed, while the original residual links connect wide layers, i.e., layers with a large number of channels; the skip connections in MobileNetV2 link the narrow layers at the begin and the end of the each block. Moreover, nonlinearities are used only in the middle of each block, while the last convolutional layer is not followed by any activation function, making the network more efficient. Xception [11, 51] also uses residual connection like ResNet and depthwise separable convolution to make the convolution more efficient but has a more complex architecture organized in flows: (i) the entry flow composed of two convolutional layers with ReLU activations and three blocks of separable convolution layers with ReLU activations and a max pooling at the end followed by a residual connection to the beginning of the block; (ii) the middle flow where blocks of three separable convolution layers with ReLU are repeated eight times and back connected by residual links; and (iii) the exit flow with a final block of separable convolution layers with ReLU and residual connection followed by a global average polling and a fully connected layer.

3 Experiments

Since our main purpose is to evaluate the suitability of machine learning methods as base to realize malware classification systems, the setup of the experimental framework has required (i) the preparation of a dataset that contains both malware and benign samples, (ii) the training of the selected methods, and (iii) the generation of the perturbed samples used to evaluate the robustness. The analysis has been conducted by firstly evaluating the base accuracy achieved by all the methods on the dataset and then analyzing the drop of accuracy related to each perturbation at different levels of severity.

3.1 Malware Dataset

The main issue while selecting a dataset to benchmark malware classification systems is the absence of benign executable samples and the poor availability of updated malware samples; all of the publicly available datasets do not deliver benign executables due to copyright restrictions. Therefore, to properly conduct our analysis, we started from a recent public dataset, namely, SOREL-20M [21], released by Sophos ReversingLabs, the AI team of Sophos, and we have extended it by adding 6899 benign samples collected from public online repositories of freeware, i.e., software freely available. The original SOREL-20M is composed of 20 million samples, 10 million malware, and 10 million benign software samples; but the latter is delivered as feature vectors only, while the executables are not available. The malware samples are disarmed to prevent accidental execution (the OptionalHeader.Subsystem flag and the FileHeader.Machine header value are both set to 0) and belong to 11 families: adware, flooder, ransomware, dropper, spyware, packed, crypto miner, file infector, installer, worm, and downloader. All the malware samples are delivered as Windows Portable Executable together with PE metadata and the features in the EMBER format as described in Sect. 2.1. For the sake of clearness, an overview of PE binary is shown in Fig. 4.

We have not used all the 10 million samples of dataset, but we have randomly selected 6899 samples from each malware family so as to obtain a balanced number of samples for each class. Therefore, the extended dataset is composed of 82.788 binaries in PE format, equally distributed over 12 classes (6899 binaries for class): the 11 original malware families of SOREL-20M with the addition of the benign class. The dataset has been divided as follows: 57,948 samples (70%) for the training set, 12,408 samples (15%) in the validation set, and 12,432 samples (15%) as test set. In Fig. 3 we show some examples of images extracted from samples belonging to the dataset, one for each class.

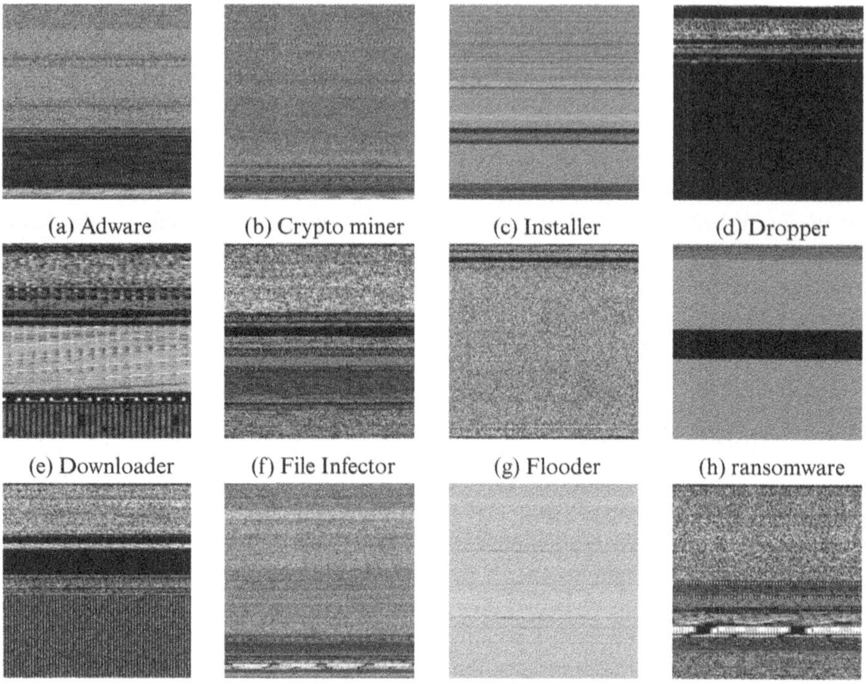

(a) Adware (b) Crypto miner (c) Installer (d) Dropper
(e) Downloader (f) File Infector (g) Flooder (h) ransomware
(i) Spyware (j) Worm (k) Packed (l) Benign

Fig. 3 Images extracted from samples in the extend SOREL-20M dataset

3.2 Generating Obfuscated Samples

Malware classification methods based on machine learning are very easy to fool, as discussed in [29]. Even small changes to the malware binary, made without affecting the functionality and behavior of the program, can be very effective in compromising the usefulness of a malware detection system. In particular, in [29] the authors append a properly generated sequence of bytes (binary padding) to the end of the executable to fool a binary classifier. In our analysis we have adopted the same approach applied to a multiclass problem; however, instead of generating an optimized sequence of bytes, we have considered simple binary paddings that can be effortlessly generated by a malicious user without any kind of knowledge about the antimalware system. We consider such paddings as baseline evasion approaches to be taken into account before extending the analysis to more sophisticated ones. Indeed, to the best of our knowledge, it has not been demonstrated if also basic patterns could significantly affect the usefulness of machine learning methods while used for malware classification. The purpose of the proposed analysis is to evaluate if the considered methods are robust against simple paddings and to compare accuracy and robustness of traditional and image-based machine learning

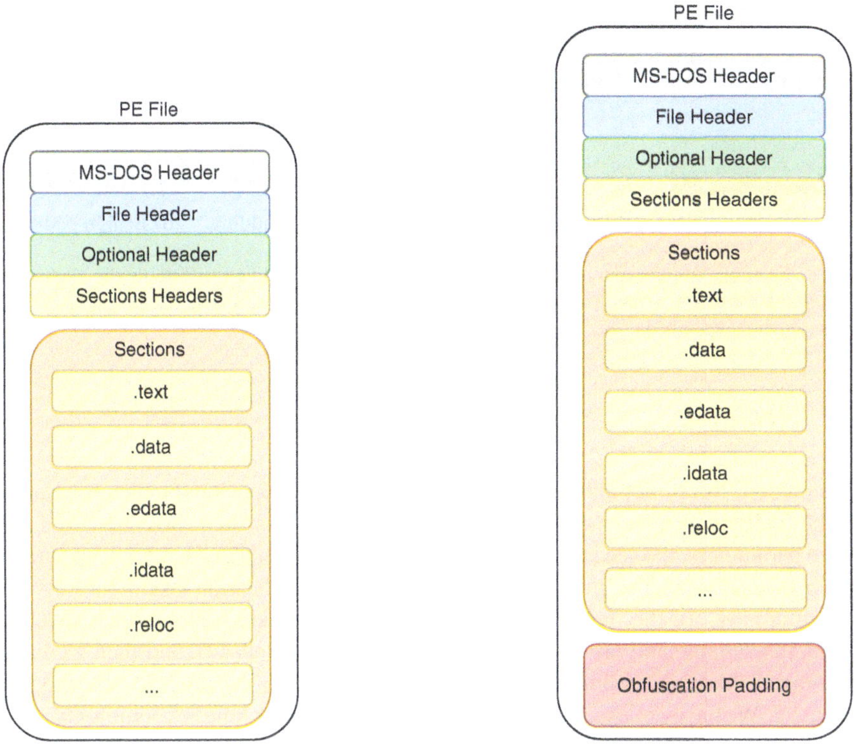

(a) Original PE file structure

(b) Obfuscated PE file structure

Fig. 4 (**a**) The common structure of a Windows Portable Executable (PE) file, composed of four main headers: (i) DOS Header, usually needed to check if the file is valid; (ii) File Header, that contains information about the file such as architecture, signatures, timestamps, size of the optional header, and pointer to the symbol table; (iii) Optional Header, containing additional information about the file like versions, checksum, DLL, file alignment, and so on; (iv) Section Header or Section Table, used to describe which sections are in the executable, where each section begins and its size. The main sections are text (executable code), data (.data, .rdata, .bss), rsrc (resources used by the file), .edata (export data), .idata (import data), and optionally the debug section. (**b**) The file after the addition of the obfuscation padding; the addition at the end has the advantage of not requiring a modification of the original headers and sections

approaches. To this purpose, we have considered the following kinds of padding: (i) *zero padding* that consists of the addition of a sequence of null bytes to the file; (ii) *random distribution*, where a random byte pattern has been generated using a uniform distribution; (iii) *junk code padding*, where the byte pattern is composed of byte sequences chosen at random from a precomputed set of byte instructions that have no effect on the functionality of the executable; and (iv) *benign padding*, where the byte pattern is composed of byte sequences extracted from benign files chosen at random. In Fig. 5 we show an example of the four different kinds of padding.

(a) Zero padding (b) Random padding (c) Junk code padding (d) Benign padding

Fig. 5 Example of the four different kinds of obfuscation padding added to the end of a packed malware. The examples show the images generated from the obfuscated adware using the representation described in Sect. 2.2

3.3 Training

The random forest methods have been trained from scratch using the feature vectors in the EMBER format as described in Sect. 2.1. For all of them, we have used the same parameters: the seeds have been chosen randomly and the categorical accuracy has been used as metric since we are considering a classification task.

Differently, for the image-based methods, we have taken CNN models that are publicly available and pre-trained on the ImageNet dataset [44], and then we have fine-tuned the CNNs using the images extracted through the process described in Sect. 2.2. We have not used any of the data augmentation techniques commonly used while training CNNs, since the image is an intermediate representation of the input binary file; the image-based data augmentation transformations would not produce a result that can be meaningfully interpreted as a valid executable. The CNNs have been trained using the categorical cross-entropy, a batch size of 32 samples, and the early stopping set to 10 epochs. For all of them, we have used learning rate with reduction on plateau; if the validation loss does not improve for three epochs, the learning rate is divided by 2 because the loss is considered to be on a plateau. We have followed the original author's recommendations for the choice of the optimizer and of the initial learning rate. In particular, we have used the stochastic gradient descent (SGD) with a learning rate of 0.1 to train Xception and MobilenetV2 and a learning rate of 0.0001 to train VGG16; we have used the Adam optimizer with a learning rate of 0.0001 to train ResNet50.

3.4 Results

The selected methods have been compared considering the original accuracy evaluated over the unobfuscated samples and the accuracy achieved when the obfuscation is applied. For each kind of padding, we have considered three levels of severity: 1%, 10%, and 25%, where the percentage represents the number of additional bytes with respect to the original size of the binary file.

As shown in Table 2, all the methods achieve a categorical accuracy higher than the 90%, so they all are able to reliably distinguish among the different families of malware and between malware and benign samples. Nevertheless, when a padding is added to the executable, even for a trivial sequence of bytes like the zero one and for a very low severity, it is possible to note a considerable drop of accuracy. For instance, the 99% of accuracy achieved by the basic random forest drops to 65% when a simple zero padding of 1% is added at the end of the file, a similar behavior can be noted also on the other methods based on random forest, i.e., XGBoost and LightGBM, even if with a lower loss of accuracy. Therefore, the first result that is evident from Table 2 is that even if the base accuracy is lower than the methods based on hand-crafted features, CNNs have autonomously learned features that are more robust with respect to small perturbations of the input for all of the considered paddings. When the severity of the perturbations grows (10% and 25%), only LightGBM and MobileNetV2 are robust against all the paddings.

It is important to note that the drop of accuracy reported in Table 2 is the average value over all the classes; thus, in order to obtaining a better understanding of each method with respect to the perturbations, it is required to analyze the confusion matrix. For the sake of space in Figs. 6, 7, and 8, we report only some noteworthy examples of the confusion matrices to provide evidences of relevant outcomes. In more details, we have to distinguish three main cases: (i) the drop is related to the confusion among malware families, but the method is still able to discriminate between malware and benign executables (see Fig. 6); (ii) the drop is widespread to all the classes, but the method can be still considered reliable (see Fig. 7); and (iii) the most parts of the malware samples are classified as benign (see Fig. 8). Of course, the case (i) is the most interesting, because the method can still be used to reliably detect malware, but not to have a classification if needed to understand the action to be taken. In case (ii) the usefulness of the method is strongly related to how it is reliable to distinguish between malicious and benign software and how much accurate is the overall prediction on other families of interest. In the last case, the method can be considered not enough robust against the addition of simple sequence of bytes and therefore completely unreliable for the task at hand. Since our focus has been to evaluate the robustness of the methods without any further training or augmentation, we have not tuned them by adding obfuscated samples in training set to evaluate how it could affect the robustness.

Table 2 The table shows the categorical accuracy of the considered methods for unobfuscated samples and for samples obfuscated with each considered combination of padding and severity. The categorical accuracy is computed as the percentage of samples where the predicted class matches with the actual one

		Zero padding			Random padding			Junk code padding			Benign code padding		
	Unobfusc.	1%	10%	25%	1%	10%	25%	1%	10%	25%	1%	10%	25%
Random forest	**0.99**	0.65	0.24	0.19	0.90	0.82	0.68	0.90	0.63	0.17	0.74	0.49	0.38
XGBoost	0.91	0.75	0.28	0.26	0.90	0.83	0.67	0.90	0.49	0.13	0.76	0.49	0.38
LightGBM	**0.99**	0.81	0.29	0.25	**0.98**	0.94	**0.79**	**0.98**	0.78	0.19	0.85	0.56	0.45
MobileNetV2	0.96	0.92	**0.79**	**0.64**	0.96	**0.96**	0.65	0.96	**0.95**	0.62	**0.95**	**0.78**	**0.60**
ResNet50	0.95	0.92	0.56	0.18	0.95	0.76	0.66	0.95	**0.95**	**0.63**	0.94	0.73	0.57
VGG16	0.94	0.81	0.22	0.09	0.94	0.63	0.41	0.94	0.93	0.16	0.93	0.44	0.26
Xception	0.95	**0.95**	0.64	0.17	0.95	0.75	0.55	0.95	**0.95**	0.48	0.95	0.73	0.49

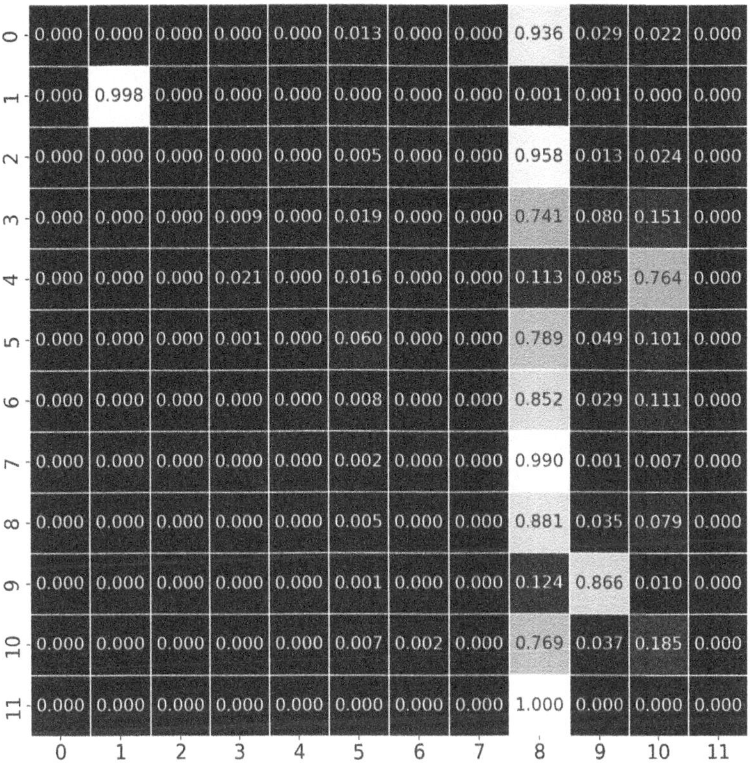

Fig. 6 Confusion matrix of LightGBM when the zero padding is added with severity 25%. In this case the LightGBM still able to distinguish between malware and benign executables, but most of families are classified as ransomware

4 Conclusions

In this paper we have proposed an initial analysis on the use of machine learning methods to deal with malware classification from executable files. The analysis has been conducted considering three traditional methods using hand-crafted features and four convolutional neural networks, with the aim of evaluating both the accuracy and the robustness against baseline obfuscation techniques based on the addition of a sequence of bytes at the end of the executable. Despite all of the methods achieved a very high categorical accuracy on unmodified test samples, we have shown that only two of them, i.e., LightGBM and MobilenetV2, are reliable enough also in case of a severe obfuscation. A future analysis will consider the addition of new methods, more complex obfuscation techniques, and evaluation of approaches to improve the robustness of the considered methods, such as the adversarial training performed by adding obfuscated samples in the training set.

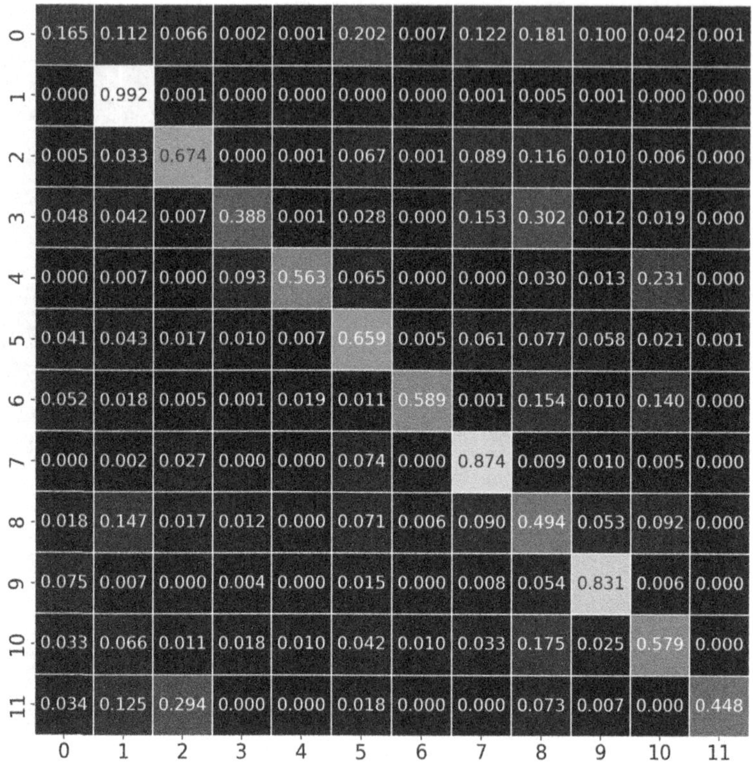

Fig. 7 Confusion matrix of MobileNetV2 when the benign code padding is added with severity 25%. In this case the categorical accuracy is 60%, but the method is still able to distinguish between malware and benign executables

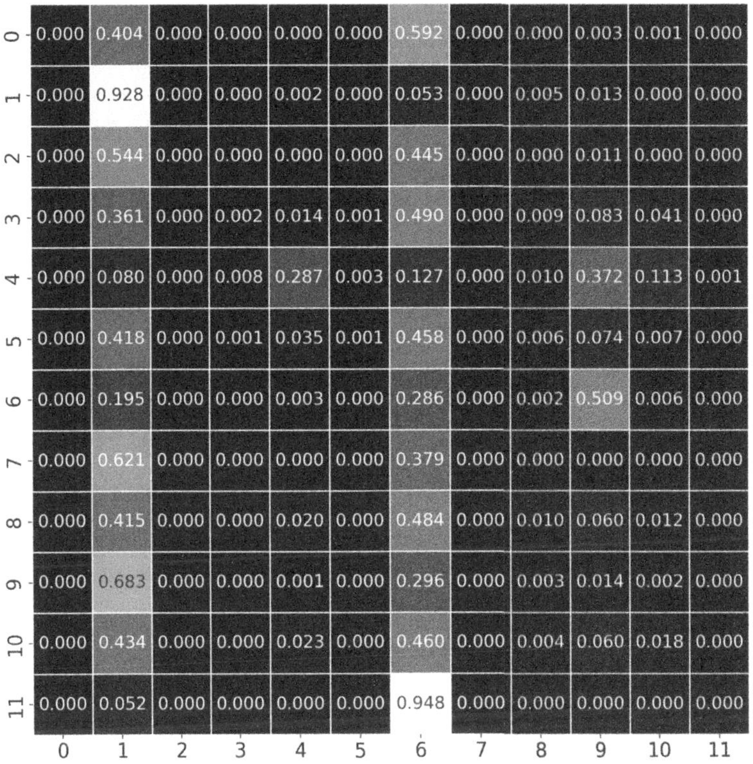

Fig. 8 Confusion matrix of XGBoost when the junk code padding is added with severity 25%. The method is useless; indeed most of the malware is classified as benign

References

1. Alam MS, Vuong ST (2013) Random forest classification for detecting android malware. In: 2013 IEEE international conference on green computing and communications and IEEE Internet of Things and IEEE cyber, physical and social computing, IEEE, pp 663–669
2. American S (2001) When did the term computer virus arise? https://www.scientificamerican.com/article/when-did-the-term-compute/
3. Anderson HS, Roth P (2018) Ember: An open dataset for training static pc malware machine learning models. 1804.04637
4. Aslan O, Samet R (2020) A Comprehensive Review on Malware Detection Approaches. IEEE Access 8:6249–6271, https://doi.org/10.1109/ACCESS.2019.2963724
5. Biggio B, Roli F (2018) Wild patterns: Ten years after the rise of adversarial machine learning. Pattern Recognition 84:317–331, https://doi.org/10.1016/j.patcog.2018.07.023
6. Bozkir AS, Cankaya AO, Aydos M (2019) Utilization and comparision of convolutional neural networks in malware recognition. In: 2019 27th Signal Processing and Communications Applications Conference (SIU), pp 1–4, https://doi.org/10.1109/SIU.2019.8806511
7. Burnaev E, Smolyakov D (2016) One-class svm with privileged information and its application to malware detection. In: 2016 IEEE 16th International Conference on Data Mining Workshops (ICDMW), pp 273–280, https://doi.org/10.1109/ICDMW.2016.0046

8. Carletti V, Greco A, Saggese A, Vento M (2021) Robustness evaluation of convolutional neural networks for malware classification. In: Proceedings of the Italian Conference on Cybersecurity (ITASEC 2021), URL http://ceur-ws.org/Vol-2940/paper35.pdf
9. Chen T, Guestrin C (2016) Xgboost: A scalable tree boosting system. In: Proceedings of the 22nd acm sigkdd international conference on knowledge discovery and data mining, pp 785–794
10. Chen T, Guestrin C (2016) Xgboost: A scalable tree boosting system. In: Proceedings of the 22nd ACM SIGKDD International Conference on Knowledge Discovery and Data Mining, Association for Computing Machinery, https://doi.org/10.1145/2939672.2939785
11. Chollet F (2017) Xception: Deep learning with depthwise separable convolutions. In: 2017 IEEE Conference on Computer Vision and Pattern Recognition (CVPR), pp 1800–1807, https://doi.org/10.1109/CVPR.2017.195
12. Corporation S (2020) Symantec internet security threat report
13. Darwaish A, Naït-Abdesselam F (2020) Rgb-based android malware detection and classification using convolutional neural network. In: GLOBECOM 2020 - 2020 IEEE Global Communications Conference, pp 1–6, https://doi.org/10.1109/GLOBECOM42002.2020.9348206
14. Davidson R (2021) The fight against malware as a service. Network Security 2021:7–11, https://doi.org/10.1016/S1353-4858(21)00088-X
15. Demetrio L, Biggio B, Lagorio G, Roli F, Armando A (2021) Functionality-preserving black-box optimization of adversarial windows malware. IEEE Transactions on Information Forensics and Security 16:3469–3478, https://doi.org/10.1109/TIFS.2021.3082330
16. Freund Y, Schapire R, Abe N (1999) A short introduction to boosting. Journal-Japanese Society For Artificial Intelligence 14(771–780):1612
17. Friedman JH (2002) Stochastic gradient boosting. Computational statistics & data analysis 38(4):367–378
18. Galen C, Steele R (2021) Empirical measurement of performance maintenance of gradient boosted decision tree models for malware detection. In: 2021 International Conference on Artificial Intelligence in Information and Communication (ICAIIC), p 93–198, https://doi.org/10.1109/ICAIIC51459.2021.9415220
19. Gibert D, Mateu C, Planes J, Vicens R (2019) Using convolutional neural networks for classification of malware represented as images. J comput virol hacking tech 15
20. GmbH AO (2020) Malware threat report:q2 2020. URL https://www.avira.com/en/blog/malware-threat-report-q2-2020-statistics-and-trends
21. Harang R, Rudd EM (2020) Sorel-20m: A large scale benchmark dataset for malicious pe detection. 2012.07634
22. He K, Zhang X, Ren S, Sun J (2016) Deep residual learning for image recognition. In: 2016 IEEE Conference on Computer Vision and Pattern Recognition (CVPR), pp 770–778, https://doi.org/10.1109/CVPR.2016.90
23. Ho TK (1995) Random decision forests. In: Proceedings of 3rd international conference on document analysis and recognition, IEEE, vol 1, pp 278–282
24. Howard AG, Zhu M, Chen B, Kalenichenko D, Wang W, Weyand T, Andreetto M, Adam H (2017) Mobilenets: Efficient convolutional neural networks for mobile vision applications. 1704.04861
25. Kabanga EK, Kim CH (2018) Malware Images Classification Using Convolutional Neural Network. Journal of Computer and Communications 06:153–158, https://doi.org/10.4236/jcc.2018.61016
26. Ke G, Meng Q, Finley T, Wang T, Chen W, Ma W, Ye Q, Liu TY (2017) Lightgbm: A highly efficient gradient boosting decision tree. In: Advances in Neural Information Processing Systems, vol 30
27. Khan RU, Zhang X, Kumar R, Tariq HA (2017) Analysis of resnet model for malicious code detection. In: 2017 14th International Computer Conference on Wavelet Active Media Technology and Information Processing (ICCWAMTIP), pp 239–242, https://doi.org/10.1109/ICCWAMTIP.2017.8301487

28. Khan RU, Zhang X, Kumar R (2019) Analysis of ResNet and GoogleNet models for malware detection. J comput virol hacking tech 15(1)
29. Kolosnjaji B, Demontis A, Biggio B, Maiorca D, Giacinto G, Eckert C, Roli F (2018) Adversarial malware binaries: Evading deep learning for malware detection in executables. In: 2018 26th European Signal Processing Conference (EUSIPCO), pp 533–537, https://doi.org/10.23919/EUSIPCO.2018.8553214
30. Le Q, Boydell O, Mac Namee B, Scanlon M (2018) Deep learning at the shallow end: Malware classification for non-domain experts. Digital Investigation 26:S118–S126, https://doi.org/10.1016/j.diin.2018.04.024
31. Li W, Ge J, Dai G (2015) Detecting malware for android platform: An svm-based approach. In: 2015 IEEE 2nd International Conference on Cyber Security and Cloud Computing, pp 464–469, https://doi.org/10.1109/CSCloud.2015.50
32. Lin Y, Chang X (2021) Towards interpretable ensemble learning for image-based malware detection. arXiv preprint arXiv:210104889
33. Lo WW, Yang X, Wang Y (2019) An Xception convolutional neural network for malware classification with transfer learning. In: 2019 10th IFIP International Conference on New Technologies, Mobility and Security (NTMS), pp 1–5, https://doi.org/10.1109/NTMS.2019.8763852
34. Mills A, Spyridopoulos T, Legg P (2019) Efficient and interpretable real-time malware detection using random-forest. In: 2019 International Conference on Cyber Situational Awareness, Data Analytics And Assessment (Cyber SA), https://doi.org/10.1109/CyberSA.2019.8899533
35. Morales-Molina CD, Santamaria-Guerrero D, Sanchez-Perez G, Perez-Meana H, Hernandez-Suarez A (2018) Methodology for malware classification using a random forest classifier. In: 2018 IEEE International Autumn Meeting on Power, Electronics and Computing (ROPEC), https://doi.org/10.1109/ROPEC.2018.8661441
36. Nataraj L, Karthikeyan S, Jacob G, Manjunath BS (2011) Malware images: Visualization and automatic classification. In: Proceedings of the 8th International Symposium on Visualization for Cyber Security, Association for Computing Machinery, https://doi.org/10.1145/2016904.2016908
37. Nisa M, Shah J, Kanwal S, Raza M, Khan M, Damasevicius R, Blazauskas T (2020) Hybrid malware classification method using segmentation-based fractal texture analysis and deep convolution neural network features. Applied Sciences 10, https://doi.org/10.3390/app10144966
38. Raff E, Barker J, Sylvester J, Brandon R, Catanzaro B, Nicholas C (2018) Malware detection by eating a whole exe
39. Raff E, Zak R, Cox R, Sylvester J, Yacci P, Ward R, Tracy A, McLean M, Nicholas C (2018) An investigation of byte n-gram features for malware classification. J comput virol hacking tech 14
40. Ramadhan FH, Suryani V, Mandala S (2021) Analysis study of malware classification portable executable using hybrid machine learning. In: 2021 International Conference on Intelligent Cybernetics Technology Applications (ICICyTA), pp 86–91, https://doi.org/10.1109/ICICyTA53712.2021.9689130
41. Rezende E, Ruppert G, Carvalho T, Ramos F, de Geus P (2017) Malicious software classification using transfer learning of resnet-50 deep neural network. In: 2017 16th IEEE International Conference on Machine Learning and Applications (ICMLA), pp 1011–1014, https://doi.org/10.1109/ICMLA.2017.00-19
42. Rezende E, Ruppert G, Carvalho T, Theophilo A, Ramos F, Geus Pd (2018) Malicious software classification using VGG16 deep neural network's bottleneck features. In: Advances in Intelligent Systems and Computing, Advances in intelligent systems and computing, Springer International Publishing, pp 51–59
43. Ronen R, Radu M, Feuerstein C, Yom-Tov E, Ahmadi M (2018) Microsoft malware classification challenge. arXiv preprint arXiv:180210135
44. Russakovsky O, Deng J, Su H, Krause J, Satheesh S, Ma S, Huang Z, Karpathy A, Khosla A, Bernstein M, Berg AC, Fei-Fei L (2015) ImageNet Large Scale Visual Recognition Challenge. International Journal of Computer Vision (IJCV) 115(3):211–252, https://doi.org/10.1007/s11263-015-0816-y

45. Saadat S, Joseph Raymond V (2021) Malware classification using CNN-XGBoost model. In: Artificial Intelligence Techniques for Advanced Computing Applications, Lecture notes in networks and systems, Springer Singapore
46. Sandler M, Howard A, Zhu M, Zhmoginov A, Chen LC (2018) Mobilenetv2: Inverted residuals and linear bottlenecks. In: 2018 IEEE/CVF Conference on Computer Vision and Pattern Recognition, pp 4510–4520, https://doi.org/10.1109/CVPR.2018.00474
47. Shafin SS, Ahmed MM, Pranto MA, Chowdhury A (2021) Detection of android malware using tree-based ensemble stacking model. In: 2021 IEEE Asia-Pacific Conference on Computer Science and Data Engineering (CSDE), pp 1–6, https://doi.org/10.1109/CSDE53843.2021.9718396
48. Simonyan K, Zisserman A (2015) Very deep convolutional networks for large-scale image recognition. In: 3rd International Conference on Learning Representations
49. Su J, Vargas DV, Prasad S, Sgandurra D, Feng Y, Sakurai K (2018) Lightweight Classification of IoT Malware based on Image Recognition. arXiv:180203714 [cs]
50. Swessi D, Idoudi H (2022) A Survey on Internet-of-Things Security: Threats and Emerging Countermeasures. Wireless Personal Communications https://doi.org/10.1007/s11277-021-09420-0
51. Szegedy C, Vanhoucke V, Ioffe S, Shlens J, Wojna Z (2016) Rethinking the inception architecture for computer vision. In: Proceedings of the IEEE conference on computer vision and pattern recognition, pp 2818–2826
52. Trend micro threat report (2019) review, refocus and recalibrate, the 2019 mobile threat landscape. URL https://www.trendmicro.com/vinfo/hk-en/security/research-and-analysis/threat-reports/roundup/review-refocus-and-recalibrate-the-2019-mobile-threat-landscap
53. Vasan D, Alazab M, Wassan S, Naeem H, Safaei B, Zheng Q (2020) IMCFN: Image-based malware classification using fine-tuned convolutional neural network architecture. Computer Networks 171, https://doi.org/10.1016/j.comnet.2020.107138
54. Vasan D, Alazab M, Wassan S, Naeem H, Safaei B, Zheng Q (2020) Imcfn: Image-based malware classification using fine-tuned convolutional neural network architecture. Computer Networks 171:107138, https://doi.org/10.1016/j.comnet.2020.107138
55. Wang G, Liu Z (2020) Android malware detection model based on LightGBM. In: Recent Trends in Intelligent Computing, Communication and Devices, Advances in intelligent systems and computing, Springer Singapore, Singapore
56. Wang J, Li B, Zeng Y (2017) Xgboost-based android malware detection. In: 2017 13th International Conference on Computational Intelligence and Security (CIS), pp 268–272, https://doi.org/10.1109/CIS.2017.00065
57. Yadav B, Tokekar S (2021) Recent innovations and comparison of deep learning techniques in malware classification : A review. International Journal on Information Security Science 9:230–247
58. Yue S (2017) Imbalanced Malware Images Classification: a CNN based Approach. arXiv:170808042 [cs, stat]
59. Zhang X, Sun M, Wang J, Wang J (2020) Malware detection based on opcode sequence and ResNet. In: Security with Intelligent Computing and Big-data Services, Advances in intelligent systems and computing, Springer International Publishing
60. Zhang Y, Li H, Zheng Y, Yao S, Jiang J (2021) Enhanced DNNs for malware classification with GAN-based adversarial training. J comput virol hacking tech 17

Formation of Reliable Composite Teams for Collaborative Environmental Surveillance of Ecosystems

Giancarlo Fortino, Lidia Fotia, Fabrizio Messina, Domenico Rosaci, Giuseppe M. L. Sarnè, and Claudio Savaglio

1 Introduction

In recent years, a significant environmental sensibility has grown throughout the world, making people aware about the importance of knowing and monitoring the state of the environment [2, 11]. This has also stimulated, on the one hand, the search for greener solutions to new and existing human processes having significant negative impacts on the environment and, on the other hand, solutions to remedy, where possible, the damage that anthropic development has caused in various ecosystems [33].

In this context, a very promising technology to change many aspects of our daily lives, making "smart" and "interactive" most of the world around us, is represented by the Internet of Things (IoT) [50]. The IoT allows to smart devices and humans to be involved together in pervasive, proactive, and collaborative activities

G. Fortino · C. Savaglio (✉)
Department DIMES, University of Calabria, Rende, Italy
e-mail: giancarlo.fortino@unical.it; csavaglio@dimes.unical.it

L. Fotia
Department DIEM, University of Salerno, Salerno, Italy
e-mail: lfotia@unisa.it

F. Messina
Department DMI, University of Catania, Catania, Italy
e-mail: messina@dmi.unict.it

D. Rosaci
Dipartimento DIIES, Università Mediterranea of Reggio Calabria, Reggio Calabria, Italy
e-mail: domenico.rosaci@unirc.it

G. M. L. Sarnè
Department of Psychology, University of Milan Bicocca, Milan, Italy
e-mail: giuseppe.sarne@unimib.it

L. Fotia et al. (eds.), *Security, Trust and Privacy Models, and Architectures in IoT Environments*, Internet of Things, https://doi.org/10.1007/978-3-031-21940-5_7

to orchestrate and execute increasingly complex and sophisticated tasks [27]. As a consequence of the opportunity provided by the IoT of interconnecting a massive numbers of smart objects with increasing computational, storage, communication, and power capabilities, new opportunities are available for governments, agencies, businesses, and citizens to know and monitor the current state and evolution over time of an ecosystem [35, 53].

By means of IoT devices, it is possible to gather over time a wide variety of data referred to the quality of air, water, and soil, as well as temperatures, electromagnetic and acoustic pollution, and so on. Moreover, the possibility of using fixed and/or mobile IoT devices and varying their spatial density makes it possible to transform almost any habitat into a smart one [7].

In this scenario, however, it must be taken into account that the potentially wide heterogeneity of devices implies that each of them will have its own trustworthiness in terms of performance [31]. The reasons for this are manifold, including the nature of the device itself, the existence of different types of devices suitable to perform the same type of measurement, adverse environmental conditions, insufficient battery power, etc. Therefore, forming the best temporary team of devices that can support human operators in monitoring a particular habitat can also be a challenging task [17], particularly in presence of large scenarios or with respect to time-critical instances or when it is necessary to take into account conflicting issues.

Some of the strategies usually adopted to select the members to form ad hoc, temporary teams (e.g., semantic similarities, structural similarities, integration or cohesion approaches, locality principles, etc.) may not be appropriate in our highly dynamic contexts and, particularly, in presence of heterogeneous devices (e.g., in terms of skills, autonomy, resource availability, etc.) [25, 52]. To address this issue, several solutions have been proposed in the literature relying on different properties, but to this end an increasing number of proposals consider social properties of devices as fundamental to have a reasonable high expectancy of achieving mutually satisfactory interactions between them [46].

In recent years, a property that has found increasing consensus in team formation processes is the *trustworthiness* [19]. To this end, several definitions of trustworthiness have been proposed in the literature [10, 15], generally dependent on the particular application context, the information sources and their freshness, the data type, etc. Consequently, several measures exist to evaluate the trustworthiness of an actor and for optimizing team-building processes, also in presence of very large communities [22, 26, 38].

Also in the formation of temporary teams to perform environmental surveillance activities, the trustworthiness of the actors can play a significant role in making such scenarios not only smart but also efficient and effective. To this end, this paper proposes a framework that adopts a trust-based approach for the automated formation of IoT device teams able to support human operators in environmental surveillance. This framework was tested on a simulated scenario demonstrating a measurable improvement in team composition both in terms of performance and appreciation.

The outline of the paper is as follows. Section 2 presents the scenario of interest and introduces the framework we designed. The details of the trust model and of the proposed team formation strategy are described in Sect. 3, while Sect. 4 presents the results of our simulations. Section 5 gives an overview on the related literature and, finally, in Sect. 6 some conclusions are drawn and the forthcoming researches delineated.

2 The Framework

As previously highlighted, nowadays there exists a greater environmental awareness, and citizens, media, and national and supranational authorities (e.g., UN, EU) are increasingly committed to its protection. Most of these actions require a synergy between different actors, at all involved levels, and the use of different tools to achieve the common goal of protecting our vital environment.

Advanced environmental monitoring processes also require similar collaborative approaches. They, if on the one hand, are made easier by advances in information and communication technologies and, on the other hand, require solving new challenges such as those of forming temporary teams of smart devices and fruitfully cooperation with human operators.

Collaborative mechanisms between sensors and humans within a team are a substantial improvement over simply collecting and analyzing sensor data. In fact, building ad hoc teams allows us to improve the effectiveness and efficiency of environmental surveillance activities and optimize the overall consumption of computational, memory, and energy resources.

In the scenario defined above, suppose we form a temporary team in which a human worker, who is the *team leader* (TL), and mobile IoT devices work together. The elements to take into account when forming such a team will be:

- A number m of smart IoT devices placed in the environment
- A number n of IoT devices necessary to carry out a specific surveillance activity
- A set of k criteria, or operation profile, characterizing the IoT devices, with respect to the surveillance activity (model, sensitivity, computational storage capacity and power, mobility, and so on).

For example, consider a complex natural ecosystem, in which a significant number of IoT devices (mobile and fixed) operate; the goal could be that to collect environmental information by selecting only those devices that meet specific requirements (e.g., detector sensitivity, etc.).

As a consequence, the surveillance process can be conducted at different levels of detail based on the number of IoT devices it uses, precision, and other characteristics. These devices will have the ability to communicate with each other and also cooperate in order to identify errors and malfunctions [42]. In this context,

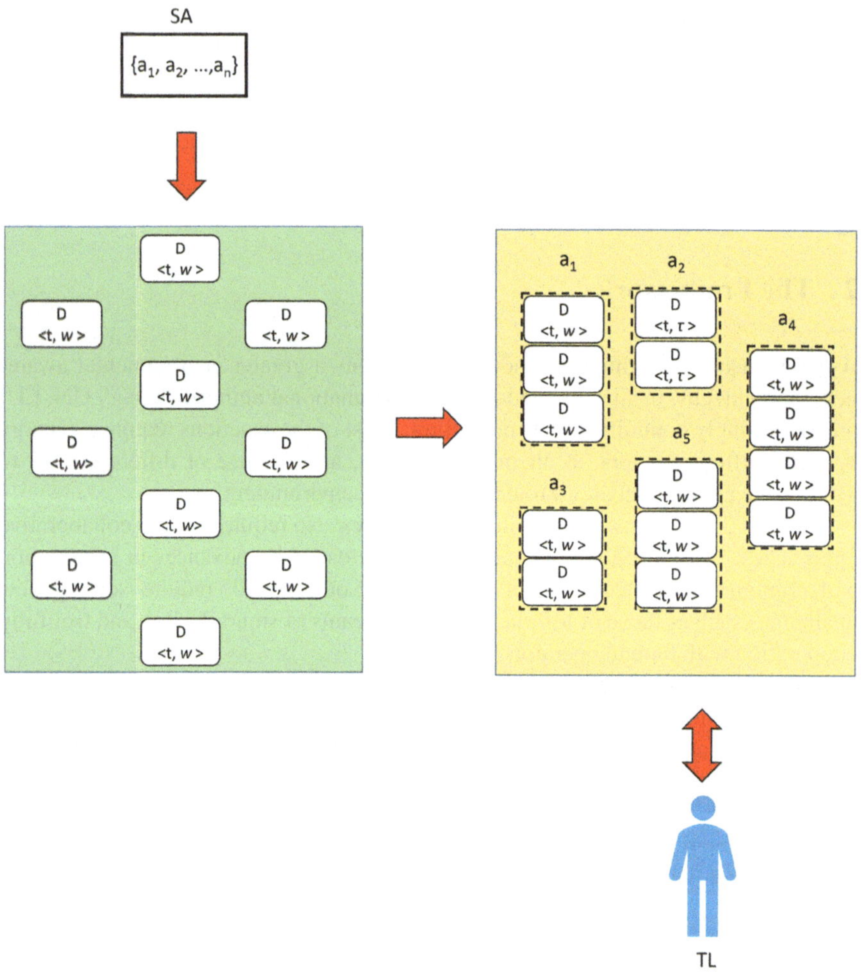

Fig. 1 The framework supports an team leader TL in forming temporary, ad hoc teams (dashed boxes) of IoT devices according to their time availability t and trust w, for carrying out some surveillance activities SA in a specific smart environment SE

the framework will be able to support the TL in the formation of the teams, analyzing and keeping the profile of the devices and the activities carried out updated. Therefore, once the task and characteristics of the IoT devices required by the TL are known, the best team of collaborators of the TL will be formed.

More formally, let A be an area to monitor, let SE the smart environment active on A, and let SA the set of the surveillance activities required to be performed on A (see Fig. 1).

Specifically, the m IoT devices are classified based on their operational profiles, which are kept constantly updated by the framework. For each SA, which can also be formed by several elementary tasks, a team leader is selected who will proceed to choose the n IoT devices members of his team T, classifying them in ascending order according to the measure:

$$R = \frac{t}{w}$$

where for each device, R is its rank, t is the operative time (e.g., the time required to a device for ending its current task and becomes again available), and w is the value of an integrated measure that takes into account the reputation and the accuracy of the IoT device (see Sect. 3).

The proposed framework is compliant with approaches that can differ in computational complexity and performance. In order to test the feasibility of the proposed solution, a numerical approach has been preliminary adopted here. In addition, a centralized or distributed approach could be adopted for managing the device profiles or for the calculation of their rank and their trust and accuracy; this also depends on the nature of the data to be processed or according to specific needs (response time, privacy, resource usage, etc.). This aspect can heavily affect the performance of the system also in the light of the specific devices features and the application context [12]. However, we currently consider these aspects as orthogonal to the core goal of our proposal.

However, it is simple to note how a collaborative and trusted monitoring activity, such as the one required in the proposed framework, can easily be implemented through the *agent-based computing* paradigm, thanks to the cooperative and social attitudes of software agents, both with real and virtual entities. In this respect, software agents can proactively perform sensing/computation/implementation tasks in the interest of human users and also applying nontrivial coordination schemes.

For this purpose, in the following we will consider that each IoT device is associated with an agent and similarly for the TL. Among the advantages of such an approach, we point out that the multi-agent system could take charge of many instances related to, e.g., device heterogeneity, thus providing the TL with an abstract, more resilient, and less critical SE.

In addition, in order to prevent tampering, protecting the integrity and reliability of the data collected and stored, the agent hosted by each device could operate autonomously from the device and appropriately encode the data (see Sect. 3) Thus, even in potential "hostile" environments, agents associated with the TL and IoT devices could mutually interact in safe modality in the absence of a central repository.

Finally, agent simulators could be exploited to simulate easily on the performance of complex coordination, computation, mobility, network, and energy models.

3 The Trust Model

In our framework, the trustworthiness of IoT devices plays a fundamental role to form the teams. To this purpose a specific trust model has been designed to characterize IoT devices. Even though a significant number of reliability, reputation and trust definition have been provided in the literature (the interested reader can refer to [3, 10, 40]). In the following we will refer to the definitions provided below:

- *Reliability*. Let k be the measure of reliability of a device, in the range [0, 1], representing its ability to correctly describe the environment where it lives. k takes into account each task the device has carried out. More in detail, let $y = \mathbf{f}(x_1, \cdots, x_n) \in [0, 1] \subset \mathrm{R}$ be a function describing the quality of the task carried out by the device and let $x_1, \cdots, x_n$ be some parameters referring to freshness, device competence, etc. Hence, k is computed as:

$$k^{new} = \alpha \cdot k^{old} + (1 - \alpha) \cdot y, \quad (1)$$

  where $\alpha \in [0, 1] \subset \mathrm{R}$ gives more or less relevance to y in updating k.
- *Reputation*. Let m be the measure of the reputation of a device, ranging in $[0, 1] \subset \mathrm{R}$, about the appreciation a device receives from the team leader community for its activity. The measure m, is computed on the basis of the feedback $n \in [0, 1] \subset \mathrm{R}$ provided by the TLs which it cooperated with and calculated as:

$$m^{new} = \beta \cdot m^{old} + (1 - \beta) \cdot n \quad (2)$$

  where $\beta \in [0, 1] \subset \mathrm{R}$ gives more or less relevance to n (i.e., the last task) in updating m.
- *Trust* Let w a measure of the trust of a device, belonging to [0, 1], which combines both *reliability* and *reputation* measures to characterize a IoT device in team formation processes. The measure w is computed as:

$$w = \gamma \cdot k + (1 - \gamma) \cdot m \quad (3)$$

  where $\gamma \in [0, 1] \subset \mathrm{R}$ gives more or less relevance to k with respect to m in order to adopt different strategies, in terms of *reliability* and *reputation*, in team formation processes based on the specific characteristics of the monitored ecosystem.

  More in detail, the main peculiarities in our reputation model are:

- *Reliability* and *reputation* respectively refers to the provided data and the data requester in order to obtain a comprehensive *trustworthiness* measure. This overcomes the syllogism that a trusted entity would, by default, guarantee trusted data. In fact, the premise itself does not guarantee that only qualitatively adequate

data can be collected, and this is without dishonest behaviors (for simplicity's sake, not currently considered) occurring for some illicit purpose.
- The proposed measure of "reliability" can take into account several contextual elements that may be present in the target functions, e.g., granularity of data (sensitivity is an important aspect in some environmental surveillance activities, like chemical radiation), frequency of detection (excessive detection frequency could represent computational, storage, and transmission overhead), responsiveness (it is critical in cases of alerts for catastrophic natural events like floods, earthquakes, and similar occurrences), etc.
- We have linearly combined *reputation* and *precision* measures in a single *trustworthiness* measure, like [23]. In such a way, by increasing the *reputation* and/or the *reliability*, then also the *trustworthiness* increases and vice versa. Notwithstanding its simplicity, behind the advantage of a predictable behavior, this linear trust model performs well, as it is shown by the results presented in Sect. 4 with respect to the proposed simulated case-study.
- Finally, the value of the γ parameter can be modified on an ad hoc basis, for example, to take into account on the contextual knowledge, or for applying more sophisticated specific strategies. To this end, sensitivity studies could be performed aimed at maximizing the performance of the framework by balancing the various factors involved therein.

4 Experiments

The proposed framework was validated through a set of experiments simulating an intelligent environment E formed by humans and IoT devices that are supported by a multi-agent system. A time window consisting of multiple working days was simulated, and a number of environmental surveillance activities were performed for each day. In turn, each surveillance task, carried out by appropriate teams of IoT devices managed by a TL, consisted of several sub-tasks, even different among them.

For simplicity, each team consists of a TL (operating as a supervisor), a number of IoT devices (for convenience assumed mobile such as simple drones). Furthermore, it was assumed that:

- IoT devices are heterogeneous in reliability and reputation (or, in other words, accuracy and effectiveness);
- Each surveillance and monitoring task requires a specialized device.

The assumptions made imply that:

- Each device requires a different amount of time to complete its task.
- A device receives from the TL that requested its service, a different appreciation.

In the formation of a team, each TL is supported by a personal agent (TLA). The TLA cooperates with the software agents associated with the IoT devices and forms, at a given time, the best possible team optimizing this process with respect to time availability and reliability of the devices (see Sects. 2 and 3). At the end of the task, the reliability (k), reputation (m), and trustworthiness (w) of the device will be updated.

4.1 Experimental Settings

In the simulation the following setting has been adopted:

- A unique smart environment SE.
- 30 working days.
- 150 surveillance (i.e., monitoring) tasks per day.
- 100 mobile heterogeneous IoT devices.
- The surveillance working place must be reached in a time of 10 min.
- Each surveillance activity consists of 4 different serial sub-tasks, each one realized by 3 different mobile IoT devices (i.e., a surveillance activity requires 12 IoT devices).

The settings of IoT devices conform to the standard values for the application domain under consideration. In addition, preliminary tests also enabled the parameters of the trust model to be configured. More in detail:

- The initial values of w and m have been initially set to 1.0, as well as that of w.
- The parameters α and β, used to calculate k and m, respectively, have both been set to 0.85, this for warranting a smooth behavior over time.
- The parameter γ, used to update the w was set to 0.75, to give more relevance to the feedback issued by the team leader.

4.2 Results

The preliminary simulations carried out used two different scenarios named $S1$ and $S2$. More specifically, two different configurations were adopted in terms of the overall reliability and reputation of IoT devices, i.e., their accuracy and their appreciation by TLs. In more detail, reliability and reputation varied in different fixed intervals in a uniform manner, and, consequently, so did the reliability measure. For the $S1$ scenario reliability and reputation varied in a range of 10%, while for $S2$ they could vary in a range of 20%.

Two strategies were also adopted for the two scenarios to guide the teams' formation processes; the first based on our approach, the other based only on the time availability of IoT devices. The results obtained are represented in Fig. 2 and confirmed that in both scenarios our strategy performs better than the one based on

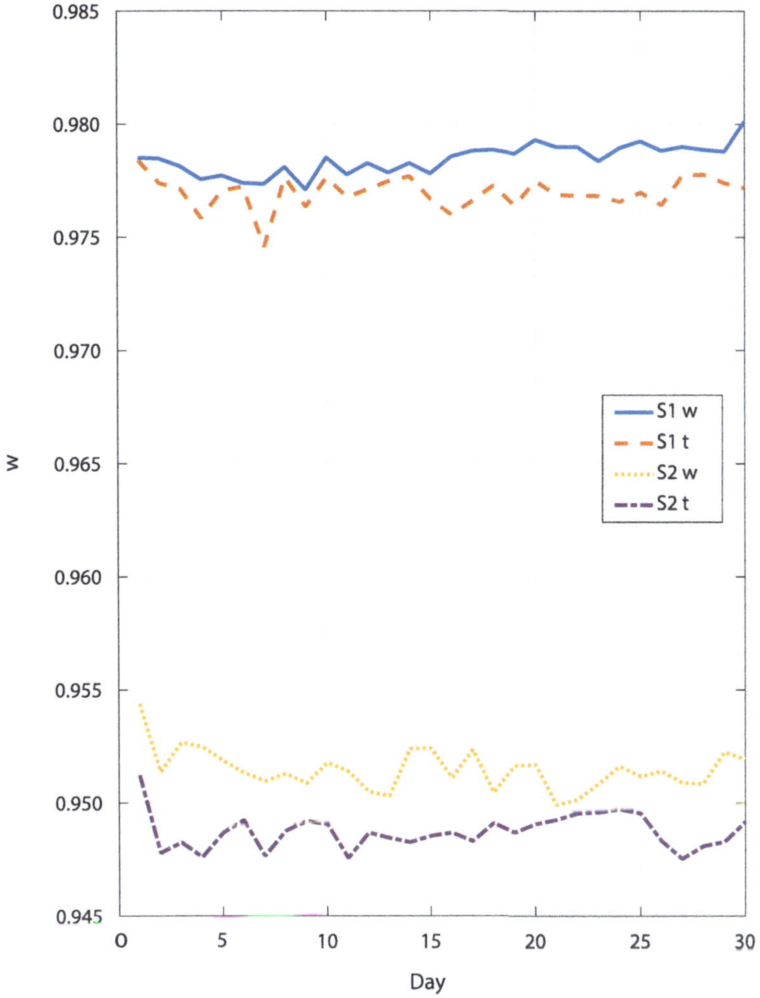

Fig. 2 Trustworthiness for working days for the scenarios S1 and S2 for teams formed on the basis of the trustworthiness and the only time, respectively

time availability alone. This implies that in the formation of the teams, those devices that are the more accurate and effective were chosen.

Furthermore, through the measures of reliability and reputation described in Sect. 3, it is possible to verify the advantages of our proposal in comparison to the formation of teams based only on the time of availability of a device or, in other words, based on the formation of the required team in the fastest way.

The two strategies described above for the two scenarios $S1$ and $S2$ were applied over the entire 30 days simulated considering the reliability and reputation measures of all IoT devices affiliated with the various teams thus formed.

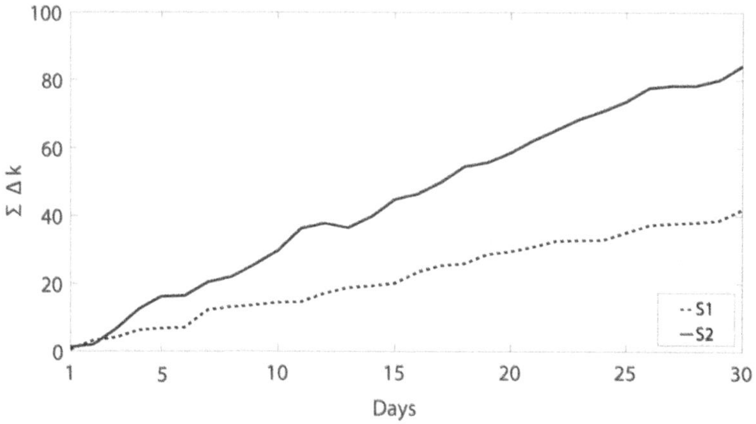

Fig. 3 Advantages in terms of reliability (k) between trustworthiness and time strategies for scenarios S1 and S2

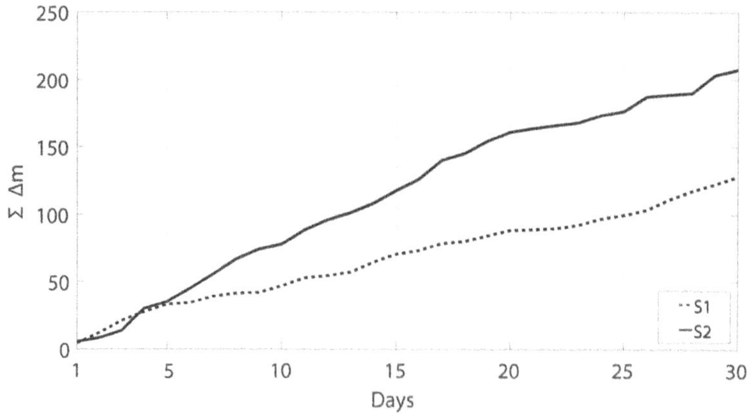

Fig. 4 Advantages in terms of reputation (m) between trustworthiness and time strategies for scenarios S1 and S2

A cumulative balance in terms of reliability and reputation for the two scenarios are depicted in Figs. 3 and 4. The benefits given by our proposal are evident in both scenarios and for both measures; moreover, the more the benefits provided by our proposal increase, the more heterogeneous the IoT devices are.

Conversely, by adopting a team formation strategy based only on the use of the devices that have maximum reliability, in that case while we will achieve the best performance in terms of accuracy if compared to any other possible strategy, but on the other hand, the time to complete a monitoring task may increase significantly and unacceptably.

5 Related Work

An increasing number of researches focus on collaborative environmental surveillance and monitoring, mainly leveraging on well-established technologies (individually considered or jointly exploited) such as wireless sensor and actuator networks (WSANs), unmanned aerial vehicles (UAVs), automated guided vehicles (AVGs), and other IoT-related ones.

In [16, 51] smart systems based on cooperating WSANs allow controlling temperature, light, and humidity levels of indoor spaces in order to identify and react to critical situations but also to improve the energy efficiency of buildings and thermal comfort of their inhabitants. In collaboration with teams of AGVs, instead, WSN-equipped workshops can benefit of reduced downtime and limited connectivity issues [20, 28]: the operations of the production lines are better synchronized with each other, while anomalies can be earlier identified and solved, often automatically, thus also limiting the human interventions.

In outdoor large-scale environments (e.g., roads, agriculture fields), WSN and UAV can collaborate in surveillance and monitoring tasks by ensuring relevant improvements in terms of effectiveness and responsiveness, especially in case of emergency situations [41, 44]. For example, collaborative smart drones have been widely exploited in the smart city domain for fire and accident detection [6], pollution [55] and parking monitoring [9], disaster management scenarios [5], and urban video-surveillance [36], often by implementing efficient data processing strategies based on lightweight AI techniques. More recently, due to the COVID-19 emergency, collaborative monitoring and surveillance systems have been implemented to hinder the virus spreading both in indoor and outdoor environments [39, 49].

In all the aforementioned works, apart from specific monitoring targets (humidity, temperature, CO/noise/pollution/water level, etc.) and implementation choices (e.g., centralized cloud-based systems or distributed ones pivoted around software agents), the collaboration between human and machines is key since the complexity of some operations demands several, heterogeneous devices to cooperate with each other and to work as one. However, to form teams by leveraging on principles such as locality [46] (i.e., to group team members geographically or socially close with each other) or similarity [13] (i.e., to ensemble team from members with common objectives, features or working modalities) is definitively challenging, especially on large-scale and dynamic scenarios. There, devices can be frequently substituted, dropped, or hijacked: therefore, the history of past interactions is not always available, and there are other few useful elements upon which estimating the reliability of candidate team members [54]. Hence, over the years, locality or similarity principles have been replaced by trust as first-class criterion for dynamic team formation, mostly supported by decentralized techniques such as EigenTrust [32] or the PeerTrust [56].

Along this line, the air-quality level and healthiness of a monitored environment has been rated in [4] by means of a trust-based decision-making protocol according to which "trusted" community members are able to forward information to health professionals and to alert people in the case of safety risks. Similarly, authors of [29] have implemented a participatory sensing for hazard detection and response system: a Trust-as-a-Service cloud utility is exploited to manage the reputation scores of the community members (i.e., users, sensors, and other IoT devices) so that they can automatically carry on both surveillance and notification tasks, even in large-scale scenarios.

Rather than scalability, trust systems of [37] and [30] put the interoperability on the spotlight. The former is SenSquare, a collaborative IoT platform for environmental monitoring capable to deal with different data formats and transmission protocols during the gathering process from heterogeneous, trusted data sources. The latter [30], instead, refers to a cloud-based data-centric framework for trustworthy IoT monitoring services: here, authors have defined a comprehensive trustworthiness score (which considers the trust of both data and entity) which is potentially compliant with the standards proposed by the Telecommunication Standardization Sector (ITU-T, which coordinates standards for telecommunications and information communication technology).

Scalability and interoperability but also smartness and autonomy are desiderata of a collaborative trust-based systems that can be achieved thanks to the software agents [24]. Just to cite a few works, an agent-based alert system has been implemented in [8] for promptly alerting in case of tsunami events: a team of "agentified" wavemeters provide certified data, automatically discarding the detection coming from less reliable or less reputed devices (according to the trust model presented by [47]).

In the industrial IoT (IIoT) scenario, instead, factory activities such as the monitoring of materials or the supply chain resource managements have been demanded to teams of trusted agents (acting as digital alias for sensors, actuators, and AVGs), automatically grouped according to mutual "on field" information/observations, social relationships, QoS, security, etc. [1, 14, 18, 21, 43, 48]. In [34] the monitoring operations of a smart grid are performed by a multi-agent system over a blockchain by exploiting game theory principles, while in [45] the authors proposed a trust-based multi-agent cooperative load balancing system aimed at monitoring CPU and memory usage reports of a highly distributed network.

In all these cases, the agent-based computing represents an enabling paradigm for the dynamic, decentralized, and scalable information processing aimed at the establishment of a sound trust system.

6 Conclusions

The formation of temporary teams that include IoT devices and human team leaders leads TLs to interact with devices whose performance, relative to their specific skills, may not be the best possible. In this article, we proposed a strategy to form the best possible team by selecting its members based on their reliability, reputation, and availability. Specific definitions of device reliability, reputation, and trustworthiness have been introduced, and a new trust-based framework was presented to support the formation of temporary teams of IoT devices.

The experiments we performed on a simulated scenario, although based on the use of realistic parameters (with the goal of considering our approach suitable for environmental surveillance applications), showed that the combination of reliability, reputation, and trustworthy information leads to measurable performance improvement in the simulations performed. In fact, the simulations have shown the potential benefits of our proposal in terms of efficiency and effectiveness.

These results obviously need to be confirmed by further experiments on real environments. Indeed, some random and unpredictable events due, for example, to malfunctions, may affect the performance of the framework. We plan to address this and other specific issues in our ongoing research.

References

1. Mohamed Abdel-Basset, Gunasekaran Manogaran, and Mai Mohamed. Internet of things (iot) and its impact on supply chain: A framework for building smart, secure and efficient systems. *Future Gener. Comput. Syst.*, 86:614–628, 2018.
2. Yandarbayeva Louisa Abdurashidovna, Adzhieva Anna Yurevna, Tsurova Liza Akhmetovna, Misakov Anzor Valerievich, and Molamusov Zalim Khashaovich. Theoretical and methodological foundations of the regional social-ecological-economic system polyfactor monitoring. *Turkish Online Journal of Design, Art and Communication*, 8:39, 2018.
3. Abdelmuttlib Ibrahim Abdalla Ahmed, Siti Hafizah Ab Hamid, Abdullah Gani, Muhammad Khurram Khan, et al. Trust and reputation for internet of things: Fundamentals, taxonomy, and open research challenges. *Journal of Network and Computer Applications*, 145:102409, 2019.
4. Hamid Al-Hamadi and Ray Chen. Trust-based decision making for health iot systems. *IEEE Internet of Things Journal*, 4(5):1408–1419, 2017.
5. Maher Aljehani and Masahiro Inoue. Safe map generation after a disaster, assisted by an unmanned aerial vehicle tracking system. *IEEJ Transactions on Electrical and Electronic Engineering*, 14(2):271–282, 2019.
6. Saeed H Alsamhi, Ou Ma, Mohammad Samar Ansari, and Faris A Almalki. Survey on collaborative smart drones and internet of things for improving smartness of smart cities. *Ieee Access*, 7:128125–128152, 2019.
7. Anwar Alshamsi, Younas Anwar, Maryam Almulla, Mouza Aldohoori, Nasser Hamad, and Mohammed Awad. Monitoring pollution: Applying iot to create a smart environment. In *2017 International Conference on Electrical and Computing Technologies and Applications (ICECTA)*, pages 1–4. IEEE, 2017.
8. Giuseppe Barbaro, Maria Donatella Gangemi, and Giandomenico Foti. An agent-based tsunami alert system. In *WOA*, pages 102–107, 2017.
9. Elizabeth Bondi, Fei Fang, Mark Hamilton, Debarun Kar, Donnabell Dmello, Jongmoo Choi, Robert Hannaford, Arvind Iyer, Lucas Joppa, Milind Tambe, et al. Spot poachers in action: Augmenting conservation drones with automatic detection in near real time. In *Proceedings of the AAAI Conference on Artificial Intelligence*, volume 32, 2018.
10. Diego De Siqueira Braga, Marco Niemann, Bernd Hellingrath, and Fernando Buarque De Lima Neto. Survey on computational trust and reputation models. *ACM Computing Surveys (CSUR)*, 51(5):1–40, 2018.

11. David E Busch and Joel C Trexler. *Monitoring ecosystems: interdisciplinary approaches for evaluating ecoregional initiatives*. Island Press, 2003.
12. Roberto Casadei, Giancarlo Fortino, Danilo Pianini, Wilma Russo, Claudio Savaglio, and Mirko Viroli. Modelling and simulation of opportunistic iot services with aggregate computing. *Future Generation Computer Systems*, 91:252–262, 2019.
13. Dong Chen, Guiran Chang, Dawei Sun, Jiajia Li, Jie Jia, and Xingwei Wang. Trm-iot: A trust management model based on fuzzy reputation for internet of things. *Comput. Sci. Inf. Syst.*, 8(4):1207–1228, 2011.
14. Ray Chen, Fenye Bao, and Jia Guo. Trust-based service management for social internet of things systems. *IEEE transactions on dependable and secure computing*, 13(6):684–696, 2015.
15. Jin-Hee Cho, Kevin Chan, and Sibel Adali. A survey on trust modeling. *ACM Computing Surveys (CSUR)*, 48(2):1–40, 2015.
16. Franco Cicirelli, Antonio Guerrieri, Alessandro Mercuri, Giandomenico Spezzano, and Andrea Vinci. Itema: A methodological approach for cognitive edge computing iot ecosystems. *Future Generation Computer Systems*, 92:189–197, 2019.
17. Philip R Cohen, Hector J Levesque, and Ira A Smith. On team formation. *Synthese Library*, pages 87–114, 1997.
18. Antonello Comi, Lidia Fotia, Fabrizio Messina, Domenico Rosaci, and Giuseppe ML Sarné. Grouptrust: Finding trust-based group structures in social communities. In *International Symposium on Intelligent and Distributed Computing*, pages 143–152. Springer, 2016.
19. Nancy W Coppola, Starr Roxanne Hiltz, and Naomi G Rotter. Building trust in virtual teams. *IEEE transactions on professional communication*, 47(2):95–104, 2004.
20. Ying Duan, Yun Luo, Wenfeng Li, Pasquale Pace, Gianluca Aloi, and Giancarlo Fortino. A collaborative task-oriented scheduling driven routing approach for industrial iot based on mobile devices. *Ad Hoc Networks*, 81:86–99, 2018.
21. Pooya Farahvash and Thomas O Boucher. A multi-agent architecture for control of agv systems. *Robotics and computer-Integrated manufacturing*, 20(6):473–483, 2004.
22. Jennifer Feitosa, Rebecca Grossman, William S Kramer, and Eduardo Salas. Measuring team trust: A critical and meta-analytical review. *Journal of Organizational Behavior*, 41(5):479–501, 2020.
23. G Fortino, F Messina, D Rosaci, and G. M. L. Sarné. Using trust and local reputation for group formation in the cloud of things. *Future Generation Computer Systems*, 89:804–815, 2018.
24. Giancarlo Fortino, Wilma Russo, and Claudio Savaglio. Simulation of agent-oriented internet of things systems.
25. Giancarlo Fortino, Claudio Savaglio, Carlos E. Palau, Jara Suarez de Puga, Maria Ganzha, Marcin Paprzycki, Miguel Montesinos, Antonio Liotta, and Miguel Llop. *Towards Multi-layer Interoperability of Heterogeneous IoT Platforms: The INTER-IoT Approach*, pages 199–232. Springer International Publishing, Cham, 2018.
26. Nicole Gillespie. Survey measures of trust in organizational contexts: An overview. *Handbook of research methods on trust*, 2015.
27. Bin Guo, Daqing Zhang, Zhu Wang, Zhiwen Yu, and Xingshe Zhou. Opportunistic iot: Exploring the harmonious interaction between human and the internet of things. *Journal of Network and Computer Applications*, 36(6):1531–1539, 2013.
28. Daqiang Guo, Ray Y Zhong, Yiming Rong, and George GQ Huang. Synchronization of shop-floor logistics and manufacturing under iiot and digital twin-enabled graduation intelligent manufacturing system. *IEEE Transactions on Cybernetics*, 2021.
29. Jia Guo, Ray Chen, Jeffrey JP Tsai, and Hamid Al-Hamadi. Trust-based iot participatory sensing for hazard detection and response. In *International Conference on Service-Oriented Computing*, pages 79–84. Springer, 2016.
30. Upul Jayasinghe, Abayomi Otebolaku, Tai-Won Um, and Gyu Myoung Lee. Data centric trust evaluation and prediction framework for iot. In *2017 ITU Kaleidoscope: Challenges for a Data-Driven Society (ITU K)*, pages 1–7. IEEE, 2017.
31. Franklin Magalhães Ribeiro Junior and Carlos Alberto Kamienski. A survey on trustworthiness for the internet of things. *IEEE Access*, 9:42493–42514, 2021.

32. Sepandar D Kamvar, Mario T Schlosser, and Hector Garcia-Molina. The eigentrust algorithm for reputation management in p2p networks. In *Proceedings of the 12th int. conf. on World Wide Web*, pages 640–651. ACM, 2003.
33. René Kemp and Babette Never. Green transition, industrial policy, and economic development. *Oxford Review of Economic Policy*, 33(1):66–84, 2017.
34. Rabiya Khalid, Omaji Samuel, Nadeem Javaid, Abdulaziz Aldegheishem, Muhammad Shafiq, and Nabil Alrajeh. A secure trust method for multi-agent system in smart grids using blockchain. *IEEE Access*, 9:59848–59859, 2021.
35. Mihai T Lazarescu. Design of a wsn platform for long-term environmental monitoring for iot applications. *IEEE Journal on emerging and selected topics in circuits and systems*, 3(1):45–54, 2013.
36. Yutong Liu, Linghe Kong, Guihai Chen, Fangqin Xu, and Zhanquan Wang. Light-weight ai and iot collaboration for surveillance video pre-processing. *Journal of Systems Architecture*, 114:101934, 2021.
37. Federico Montori, Luca Bedogni, and Luciano Bononi. A collaborative internet of things architecture for smart cities and environmental monitoring. *IEEE Internet of Things Journal*, 5(2):592–605, 2017.
38. Michael Naef and Jürgen Schupp. Measuring trust: Experiments and surveys in contrast and combination. 2009.
39. Liwei Ouyang, Yong Yuan, Yumeng Cao, and Fei-Yue Wang. A novel framework of collaborative early warning for covid-19 based on blockchain and smart contracts. *Information sciences*, 570:124–143, 2021.
40. Isaac Pinyol and Jordi Sabater-Mir. Computational trust and reputation models for open multi-agent systems: a review. *Artificial Intelligence Review*, 40(1):1–25, 2013.
41. Dan Popescu, Florin Stoican, Grigore Stamatescu, Oana Chenaru, and Loretta Ichim. A survey of collaborative uav–wsn systems for efficient monitoring. *Sensors*, 19(21):4690, 2019.
42. Maria Nadia Postorino and Giuseppe M. L. Sarné. An agent-based sensor grid to monitor urban traffic. In *Proceedings of the 15th Workshop dagli Oggetti agli Agenti, WOA 2014*, volume 1260 of *CEUR Workshop Proceedings*. CEUR-WS.org, 2014.
43. Christian Prasse, Andreas Nettstraeter, and Michael Ten Hompel. How iot will change the design and operation of logistics systems. In *2014 International Conference on the Internet of Things (IOT)*, pages 55–60. IEEE, 2014.
44. Yerra Prathyusha and Chung-Nan Lee. Uav path planning and collaborative searching for air pollution source using the particle swarm optimization. In *International Computer Symposium*, pages 698–709. Springer, 2018.
45. US Premarathne and S Rajasingham. Trust based multi-agent cooperative load balancing system (tclbs). *Future Generation Computer Systems*, 112:185–192, 2020.
46. Syama Sundar Rangapuram, Thomas Buhler, and Matthias Hein. Towards realistic team formation in social networks based on densest subgraphs. In *Proceedings of the 22nd int. conf. on World Wide Web*, pages 1077–1088. ACM, 2013.
47. Domenico Rosaci, Giuseppe M. L. Sarné, and Salvatore Garruzzo. Integrating trust measures in multiagent systems. *International Journal of Intelligent Systems*, 27(1):1–15, 2012.
48. Yosra Ben Saied, Alexis Olivereau, Djamal Zeghlache, and Maryline Laurent. Trust management system design for the internet of things: A context-aware and multi-service approach. *Computers & Security*, 39:351–365, 2013.
49. Claudio Savaglio, Giancarlo Fortino, Giandomenico Spezzano, Mario Alejandro Paguay Alvarado, Fabio Capparelli, Gianmarco Marcello, Luigi Rachiele, Francesco Raco, and Samantha Genoveva Sanchez Basantes. Edge intelligence against covid-19: a smart university campus case study. In *IoT Edge Solutions for Cognitive Buildings*, pages 1–25. Springer, 2022-to appear.
50. Sajjad Hussain Shah and Ilyas Yaqoob. A survey: Internet of things (iot) technologies, applications and challenges. In *2016 IEEE Smart Energy Grid Engineering (SEGE)*, pages 381–385. IEEE, 2016.

51. Muhammad Tahir, Rabia Noor Enam, Syed Muhammad Nabeel Mustafa, and Rukaiya Javed. Emas: Environment monitoring and smart alert system for internet of things (iot). In *2020 Global Conference on Wireless and Optical Technologies (GCWOT)*, pages 1–4. IEEE, 2020.
52. Shun Takai and Marcos Esterman. Towards a better design team formation: A review of team effectiveness models and possible measurements of design-team inputs, processes, and outputs. In *International Design Engineering Technical Conferences and Computers and Information in Engineering Conference*, volume 58158, page V003T04A018. American Society of Mechanical Engineers, 2017.
53. Silvia Liberata Ullo and GR Sinha. Advances in smart environment monitoring systems using iot and sensors. *Sensors*, 20(11):3113, 2020.
54. Ke Wang, Ship Peng Xu, Chien-Ming Chen, SK Hafizul Islam, Mohammad Mehedi Hassan, Claudio Savaglio, Pasquale Pace, and Gianluca Aloi. A trusted consensus scheme for collaborative learning in the edge ai computing domain. *IEEE Network*, 35(1):204–210, 2021.
55. Josefa Wivou, Lanka Udawatta, Ali Alshehhi, Ebrahim Alzaabi, Ahmed Albeloshi, and Saeed Alfalasi. Air quality monitoring for sustainable systems via drone based technology. In *2016 IEEE International Conference on Information and Automation for Sustainability (ICIAfS)*, pages 1–5. IEEE, 2016.
56. Li Xiong and Ling Liu. Peertrust: Supporting reputation-based trust for peer-to-peer electronic communities. *IEEE transactions on Knowledge and Data Engineering*, 16(7):843–857, 2004.

Accountability of IoT Devices

Angelo Furfaro, Carmelo Felicetti, Domenico Saccà, and Felice Crupi

1 Introduction

The worldwide diffusion of IoT devices is experiencing an exponential growth despite the chip shortage and COVID-19's impact on the supply chain. Their ubiquity and pervasiveness are acting as the main drivers for the development of new software services and applications [30]. At the same time, this huge number of Internet-connected objects, which expose online the physical world and which need to operate quickly, reliably, and to interact autonomously with other systems and devices, raises important accountability issues related to safety, privacy, and security of the data they operate on [32]. This scenario is made more complex by the fact that rapid rush to the market has often led to the neglect of cybersecurity problems due to the lack in the use of security-by-design development approaches [16].

As pointed out in [32], IoT devices are part of a *systems-of-systems* environment where heterogeneous components, governed by different (legal) entities, interact among them to carry out complex interorganizational business processes [29]. In such scenario, a very important requirement regards the accountability of the actions executed during process enactment. Hence, trusted mechanisms have to be provided that allow to trace actions and securely attribute them to the relevant involved actors, in such a way that responsibility can be claimed with an appropriate level of assurance, and, thus, actions cannot be repudiated or falsely attributed [3, 24]. Another important aspect is the need to achieve accountability without resorting to a central government authority acting as a trusted third party (TTP). Some decentralized solutions, relying on the joint use of blockchain technology [25, 26] and public digital identity systems (e.g., eIDAS [5]) have been described in

A. Furfaro (✉) · C. Felicetti · D. Saccà · F. Crupi
DIMES, University of Calabria, Rende, Italy
e-mail: angelo.furfaro@unical.it; carmelo.felicetti@unical.it; sacca@unical.it; felice.crupi@unical.it

L. Fotia et al. (eds.), *Security, Trust and Privacy Models, and Architectures in IoT Environments*, Internet of Things, https://doi.org/10.1007/978-3-031-21940-5_8

literature [2, 3, 8, 17]; however, such systems do not consider actions which may be autonomously made by IoT devices involved in the process loop.

The infrastructure here proposed takes its cue from the results summarized in [3] by adding the support for tracking interactions originated by IoT devices identified by exploiting physically unclonable functions (PUFs) technology [19], as preliminarily discussed in [15]. PUFs allow to build something similar to a digital identity for physical devices (e.g., IoT devices). This technology exploits variations occurring the manufacturing process of integrated circuits, and it is based on the extraction of some parameters (e.g., voltage, current, time delay), depending on such variations, that can be considered as the fingerprint of the chip itself. PUF-based identification of IoT devices is accomplished through a special type of blockchain *master nodes* [11], named *accountability nodes* (ANs). ANs, in turn, are in charge to operate on the blockchain for tracing the interaction of the IoT devices they are able to authenticate.

In order to enable the traceability of relevant interactions, their enactment is achieved by means of appropriate *smart contracts* [33] executed on the blockchain. When IoT devices have to perform actions subject to traceability requirements, they need first to be authenticated, by means of PUFs they are equipped with, and then they may use the accountability service offered by the infrastructure. Device authentication occurs through the brokerage of an AN. Then, the blockchain will keep track of the transaction by storing in the public ledger only the information needed for ensuring the accountability requirements.

The remainder of the chapter is structured as follows. Section 2 discusses the related work. Section 3 gives an overview of the key technologies at the basis of the proposed infrastructure. Section 4 provides a description of the infrastructure, including main actors that interact with it and its key components. Finally, Sect. 5 draws the conclusions.

2 Related Work

The pervasive diffusion of IoT devices poses several accountability challenges involving various aspects, e.g., governance and responsibility, privacy and surveillance, and safety and security, as has been highlighted in [32]. The abovementioned work gives an overview of the accountability issues from the legal dimension perspective and discusses some technical directions toward the enabling of IoT accountability. In particular, blockchain technologies are foreseen as one of the main building blocks for achieving technical auditing mechanisms.

The work published in [7] investigated how a blockchain infrastructure can be adopted for securing the deployment of updates for IoT devices. Protecting IoT devices from the injection of malicious code is a critical aspect. There are some situations where the update of the core software components running on a device, e.g., the firmware, is mandatory for ensuring its correct and safe operations. In such cases, it is very important to ensure that the updates are triggered by and come from trusted sources and that their outcomes are tracked. The proposed approach is interesting; however, it is limited to securing software updates in IoT devices and does not account neither for device identification nor for tracking other relevant interactions with external systems.

An approach aimed at achieving IoT accountability in compliance with the European *General Data Protection Regulation* (GDPR) [1] is discussed in [10], where the so-called IoT Databox model is proposed. An IoT Databox is an edge device that is intended to be situated within the home, which collates data from IoT devices, either directly or via APIs, and makes them available to "apps" that enable data processing and actuation. While interesting, this model, in addition to a wide adoption and diffusion of such edge devices, would require to build an architecture where a standard set of development tools, one or more dedicated app store, and third-party data processors are put in operation to (try to) enforce GDPR compliance of IoT applications.

Regarding the use of PUF technologies for device authentication, a PUF-based communication protocol enabling the secure authentication and communication among the IoT devices has been proposed in [9]. A public key cryptography approach relying on identity-based encryption (IBE) [6, 13] is adopted where each IoT device is paired with a PUF employed to generate the respective public identity. The authors provided formal proofs which demonstrate the resilience of the proposed scheme against passive and active attacks.

A survey of PUF security schemes along with a classification based on the exposure of the transmitted CRPs (see Sect. 3.2) has been published in [23]. The work presented in [31] gives a survey on PUF-based security solutions for IoT.

3 Key Technologies

3.1 Blockchain

One of the most challenging issues for achieving a non-disputable attribution of responsibility and/or liability for interactions occurring in a distributed cooperation environment, without relying on a trusted third party, has been for long time that of reaching a consensus among all the involved entities according to a decentralized peer-to-peer approach. This is a very old problem that has been solved with the advent of blockchain technology, recently introduced to actually implement cryptocurrencies, the most popular of which is the Bitcoin [28].

Since its introduction, blockchain technology has evolved under the driving force of the potential new fields of applications, whose requirements led in turn to the introduction of new features. A recent study [20] highlights the leading position of blockchain in today's industrial applications and discusses the main issues and challenges in the context of Industry 4.0 where IoT is the main source of innovation.

Table 1 Blockchain technologies' evolution

	Main features	Applications
1.0	Distributed ledger	Cryptocurrencies
2.0	Smart contracts	Decentralized Autonomous Organizations and Corporations (DAOs and DACs)
3.0	Governance	E-voting, E-health, notary systems
4.0	Industry 4.0	Cyber-physical systems (CPS), smart manufacturing, industrial Internet of Things (IIoT)
5.0	Artificial intelligence	Web 3.0 applications

The evolution of the support for new features and applications is here briefly sketched in Table 1.

A blockchain essentially consists of a peer-to-peer network, which operates without the control of any central authority and whose goal is to maintain a distributed ledger shared among all the entities connected by the network which in turn agree on the status of the ledger. All the participants keep an identical copy of the ledger and collaborate in order to consistently update it by adding new entries in the ledger through a certified and reliable mechanism. Ledger updates are made by means of *transactions* which are always signed by the involved entities , and for them to be added into the ledger, they should first undergo a public verification and validation process executed by the peer nodes. Because of this, each entity wanting to participate in blockchain transactions must own a public/private key pair used to uniquely identify it and to sign each transaction in which it is involved. This happens without any need to link the *virtual* identity, represented by the public key, with the physical identity in the real world. As pointed out before, an entity may correspond to either a human being or to an (autonomous) physical IoT device. The distributed consensus on the blockchain is achieved through a decentralized *proof of work* (PoW) mechanism that requires members of the network to expend effort solving an arbitrary mathematical puzzle to prevent anybody from gaming the system.

After the diffusion of cryptocurrencies, the growing interest generated by the blockchain, due to its potentials, has favored a constant and rapid evolution to support new types of applications. An important innovation was represented by the introduction of the so-called *smart contracts*, which led to the blockchain 2.0. Smart contracts [22] are simply programs stored on a blockchain that run when predetermined conditions are met. They typically are used to automate the execution of an agreement so that all participants can be immediately certain of the outcome, without any intermediary's involvement or time loss. Smart contracts enable *programming* the blockchain allowing the development and deployment of distribute applications, different from cryptocurrencies, which exploit both the consensus protocol and the distributed ledger. The Ethereum blockchain is the most famous example of programmable blockchain. Ethereum extends the notion of distributed ledger with that of a *state machine* whose state changes from block to block according to a pre-defined set of rules and which can execute arbitrary machine code [14].

The successive evolution of blockchain (3.0) introduced *master nodes* (MNs), which are special network nodes that, unlike regular nodes, do not add new blocks of transactions to the blockchain. Instead, they verify new blocks and perform special roles in governing the blockchain. They are also in charge of providing resource-intensive services and operate according to a *proof-of-stake* collateral-based system, meaning the operators need to own a significant amount of the cryptocurrency. In exchange for their investment in time and money, master node operators are rewarded with guaranteed crypto earnings, usually a percentage of their stake [11]. The activation of a MN is subordinated to a very conspicuous initial investment. The purpose of this requirement is essentially due to the fact that the fear of losing the investment steers operators to keep the network in good condition, watch over it, and prevent malicious users from trying to damage it. When one of the main nodes performs an illegal action, as soon this behavior is detected, the node is disabled, and its owner loses forever the initial investment bail.

Blockchain 4.0 refers to the diffusion of blockchain technology in the context of industrial systems for enabling the implementation of real-world decentralized industrial applications. Blockchain 5.0 foresees the integration of blockchain technology with artificial intelligence (AI) to create the next generation of decentralized Web 3.0 applications, ensuring data privacy, security, and interoperability [20].

3.2 Physical Unclonable Functions

With the technological scaling of integrated circuits (ICs) and the advent of sub nanometric technologies, it is more difficult to have the exact control on the manufacturing process of IC-based devices. These intrinsic process variations bring to fluctuation in the physical and geometric transistor parameters (e.g., channel length, oxide thickness, and doping levels), which translate into different electrical characteristics.

Such dissimilarities in the circuit microstructures, occurring during the manufacturing process of a set of otherwise identical (from the viewpoints of both design and layout) ICs, are exploited to achieve physical unclonable functions (PUFs). A PUF is an IC which reacts to an external input (*challenge*) by producing a unique unpredictable output (*response*) [23], and its behavior can be represented through a function $f : C \rightarrow R$, where C and R are the challenge and the response domains, respectively. Given a set P of N PUFs, $P = \{p_0, p_1, \ldots, p_{N-1}\}$, identically designed and built during the same process, we have that each function $f_i(\cdot)$ has a different, yet deterministic, behavior which is determined by the abovementioned variations due to the manufacturing process.

Because of their characteristics, PUFs are a promising innovative primitive which can be exploited for achieving secure authentication protocols for physical devices. The four fundamental properties making PUFs suitable to be employed in secure systems [19] are:

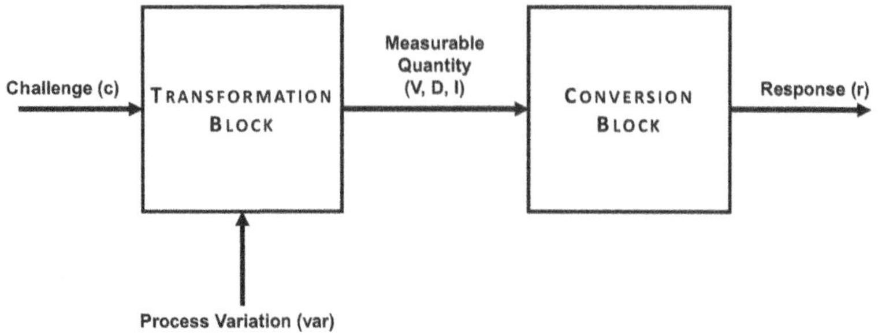

Fig. 1 Silicon-based PUF architecture [19]

- **Uniqueness**, according to which, by building different instances of the same PUF design, even during the same manufacturing process, each of them assume an individual behavior. Each PUF instance produces a different response when stimulated with the same input challenge. The uniqueness of an object is often associated with the concept of inter-distance [4].
- **Unpredictability**, which is represented by the fact that although the response to a specific challenge or to a set of previous challenges is known, it is not possible, however, to foresee the response to a challenge not yet processed by the PUF. The underlying reason is due to the process variations are so unpredictable that they cannot be modeled with a mathematical algorithm;
- **Randomness**, intrinsic in all integrated circuits, and due to the combined effect of process variations and the background noise typical of the manufacturing phase of electronic circuits. It represents the property capable of ensuring a fair balance between bits equal to 0 and 1 to internal of the strings relating to the response provided by any PUF.
- **Physical Unclonability**, it is the characteristic due to the realization process that never allows to have two instances of the same PUF design that have an equal set of CRPs.

Generally, the PUF architecture is composed by a transformation block (Fig. 1) which turns a challenge into a measurable quantity (voltage, current, or delay) combining the effect of the challenge which acts as an enabling signal, with the process variations affecting the circuit with which the transformation block itself was built, and by a conversion block which produces a binary response. The former acts as an unpredictable element, the result of which is different for each manufactured sample; the latter instead acts in a standard way, by making a simple adjustment of the signal to be read in digital form. In any case, both blocks are made through the reuse of already well-known and widely diffused circuitry, offering a practical solution to implement and not particularly burdensome in terms of costs.

The challenge-response space of a PUF determines whether it is classified as a *weak* PUF or a *strong* PUF. Strong PUFs have a large challenge-response space that grows exponentially with its implementation size; this allows their usage in

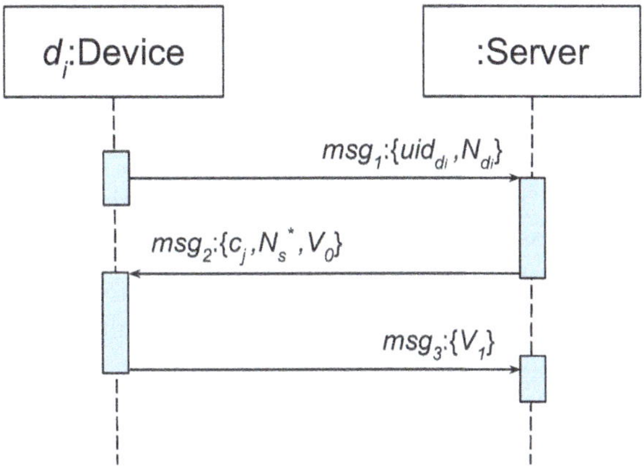

Fig. 2 PUF authentication

challenge/response authentication protocols where each challenge-response pair is used only once [21, 23]. An example of a strong PUF-based authentication protocol, which uses the challenge-response mechanism, is described in the following. A comparative study of various PUF-based security protocols, suited for IoT devices, has been recently published in [18].

The protocol relies on the presence of a trusted authentication server which, for each device d_i, safely stores a subset of its challenge-response pairs (CRPs) obtained during the so-called *enrollment* phase. To enroll a device d_i, equipped with a PUF p_i, the server first generates a set of challenges $C_i = \{c_1, c_2, \ldots, c_n\}$ and ask d_i to generate the corresponding responses $R_i = \{r_1, r_2, \ldots, r_n\}$, where $r_j = f_i(c_j)$ and $f_i(\cdot)$ are the function of PUF p_i. The server stores all the sampled CRPs for device d_i and keeps them associated with the device's unique identifier uid_{d_i}, $CRP[uid_{d_i}] = \{(c_1, r_1), (c_2, r_2), \ldots (c_n, r_n)\}$. A critical aspect regards the fact that the device enrollment phase must be accomplished in secure setting, where no other entity, apart from the server, is allowed to obtain the sampled CRPs. For this reason, such phase is typically performed by the device manufacturer which is also responsible to run and maintain the authentication server.

The authentication phase, detailed in the sequence diagram reported in Fig. 2, is accomplished as follows.

- Device d_i, which asks to be authenticated, generates a random *nonce* N_{d_i} and sends it, along with its identifier uid_{d_i}, into a message directed to the authentication server, $msg_1 : \{uid_{d_i}, N_{d_i}\}$.
- Upon receiving the authentication request message, the server looks up its CRPs database and randomly selects a pair (c_j, r_j) among those associated to uid_{d_i}. Then, it generates a random nonce N_s and computes $N_s^* = r_j \oplus N_s$ and $V_0 = h(N_{d_i} \parallel r_j \parallel N_s^*)$, where $\oplus$ denotes the bitwise exclusive or operation, h is a suitable one-way hash function, and $\parallel$ denotes message concatenation. The server sends a message $msg_2 : \{c_j, N_s^*, V_0\}$ to the device.

- When the device receive msg_2, first verifies its validity by comparing the value of V_0, contained in msg_2 with that obtained from locally computing $h(N_{d_i} \parallel r_j \parallel N_s^*)$. If validation is not successful, the process is aborted; otherwise the device extracts the server nonce $N_s = N_s^* \oplus r_j$, computes $V_1 = h(N_s \parallel r_j)$, and sends a message $msg_3 : \{V_1\}$ to the server.
- Upon receiving msg_3, the server validates the value of V_1. If validation is successful, this witnesses the identity of the device which is then authenticated by the server.

The database may be subject to attacks; for this reason every time a CRP is used, it is definitively deleted. Such operation allows to prevent man-in-the-middle attacks as the challenges are never reused, allowing the transmission of CRPs in plaintext during the authentication phases. Once all the stored CRPs are used, a refresh phase has to be performed to generate new CRPs.

4 IoT Accountability Infrastructure

The infrastructure here proposed combines the above-described key technologies to build the accountability services it provides. In order to make an IoT object responsible for the actions it performs, it must first of all be identified and authenticated, and then its relevant interactions have to be tracked. Therefore, the fulfillment of the basic criteria for the implementation of service accountability is entrusted to the features offered by PUF and blockchain technologies.

The blockchain, in addition to enabling the service decentralization, also assumes the role of guarantor, keeping track of the operations carried out and offering maximum transparency, preventing possible alterations. By making the information public and allowing anyone who is part of the network to freely verify it, the involved entities will certainly maintain a behavior that does not damage their reputation.

PUF technology, on his side, allows to establish a solid link between the physical objects operating in the IoT context and their respective digital identities which are in turn used when they interact, through the infrastructure, with other entities. The PUF-based authentication mechanism is therefore necessary for the attribution of responsibilities for the actions carried out by the IoT objects.

As discussed in more detail later, the infrastructure stacks on top of an existing blockchain 3.0 and consists of a set of organizations interested in performing accountable interactions for the execution of decentralized cooperative business processes. These organizations operate a certain number of special-purpose master nodes, known as *accountability nodes* (ANs). ANs operate on the blockchain, by executing smart contracts and relevant transactions, on behalf of entities which have been authorized by a participating organization.

The infrastructure has been devised to enable accountable interactions which may involve both legal entities, e.g., people working for companies, and IoT devices.

The former authenticate themselves through public digital identity providers, e.g., eIDAS compliant [5], the latter by means of device-specific identity providers, typically operated by device manufacturers, which exploit PUF-based authentication mechanisms to certify IoT devices identity.

The infrastructure defines two main roles: *users* and *managers*. Users are those entities which need to perform an accountable interaction, for example, during the execution of a step in a cooperative process. As pointed out before, users may be either autonomous devices or people. Managers carry out coordination activities, dynamically regulating users permissions by granting the access to specific interactions. Both users and managers, before taking any action, mediated by an AN, need to be first identified and authenticated by their respective providers. During the identification phase, the AN acts as a *service provider* [5].

In order for a user to be able to use the accountability infrastructure, it has first to be enabled by registering its digital identity with an AN operated by the organization on behalf of which it will act. This registration step is achieved by means of a specific smart contract requested by a manager working for the organization. This smart contract is actually executed by the AN which shoulders the burden of signing and verifying the actual carrying out of the relevant transactions. As a consequence, it is the organization operating the AN which is then charged for the relevant fees.

By default, when an organization subscribes to the infrastructure to take part in cooperation processes, it is not authorized to take any action. Whenever a user linked to a specific organization has to perform a certain operation, a specific role is assigned to him, and only the necessary authorizations are provided to carry out the assigned task. The burden of activating the necessary authorizations is entrusted to the managers, who fulfill their role by activating smart contracts. The release, as well as the revocation of each authorization, is therefore subject to the activation of a smart contract, the execution of which is written on the ledger in the form of transactions. The actions carried out by users and managers and the various interactions that take place among them within the infrastructure are defined by specific web services.

The following sections detail the architecture and describe the key components of the infrastructure.

4.1 Architecture

The infrastructure described is made up of a series of interconnected elements as depicted in the scheme shown in (Fig. 3).

As discussed before, the infrastructure, considering its purposes, must rely on a stable, flexible, and easily extensible blockchain 3.0. Ethereum [14], for example, has shown in the past that it has the appropriate requirements. However, even if it supports many services useful for the operation of the infrastructure, in relation to how it was conceived, the main limitation is represented by the fact that the Ethereum project was born to respond to the requirements of blockchain 2.0,

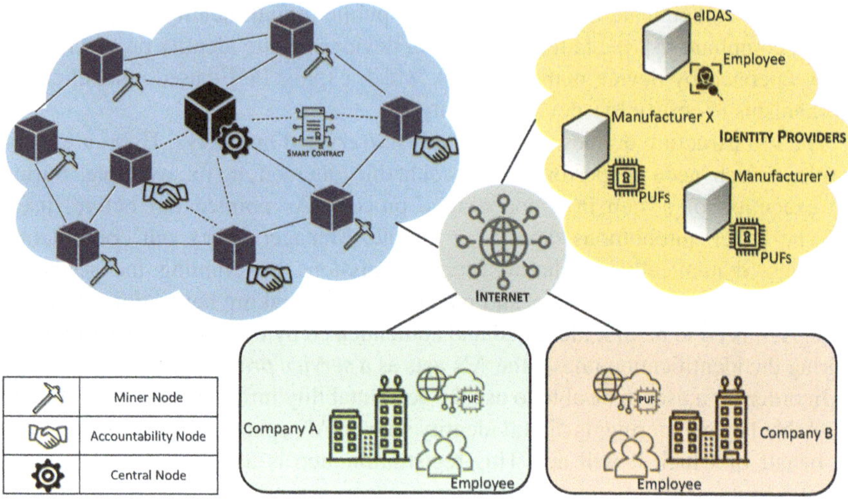

Fig. 3 Infrastructure architecture

whereas the proposal specifically makes use of technologies specific to the 3.0 blockchain. In particular, the architecture requires that the blockchain, in addition to the traditional miner nodes, includes two specific types of additional nodes, which are the accountability nodes (ANs) and the central node (CN). Both are special types of master nodes.

Therefore, either a fork of Ethereum, in the form of permissioned blockchain, has to be developed by including such type of nodes or blockchain like Dash [12] has to be adopted.

As the name suggests, ANs provide the accountability service and keeps the network in good health. The central node, on the other hand, has the fundamental task of coordinating ANs, but its role is completely transparent to the users. The central node has the commitment to verify that the registration and affiliation procedures carried out by the accountability nodes are successfully carried out and to distribute to them the rewards for their contributions in the infrastructure operations.

Finally, the miner nodes do not participate in the accountability process but nevertheless have a fundamental role, which consists in creating and introducing new blocks within the chain.

Looking beyond the blockchain, we also find the authentication providers that have a primary role in the functioning of the infrastructure, in that they ensure the identification of users, devices, or people, participating in the cooperation process, creating a link between the entity in the real world and its alter ego in the digital world, with all the consequences that it implies, even in terms of accountability.

The authentication service is offered by an independent third party, normally represented by the PUF/IoT manufacturer in the case devices, and by public identity providers for people.

The following sections present an in-depth analysis of the role and behavior of the blockchain nodes and of the other elements that make up the architecture.

4.1.1 Accountability Node

The accountability node plays a role similar to a master node, characteristic of blockchain 3.0 technology, and essentially has the purpose of providing services to the network to which it is connected.

The provision of services and the assumption of responsibility by an accountability node must be balanced by a value placed as a *guarantee* or *stake* for the commitment it assumes. Therefore, any miner node that has enough cryptocurrency to fulfill this role can become accredited as an accountability node by initiating the related registration process. The escrow deposit has the purpose of discouraging the accountability node from engaging in any behavior that may be considered unsuitable. This appears entirely motivated by the fact that it acts as a sort of intermediary among users and the blockchain.

Another important task performed by the accountability node is to act as a delegate within the blockchain on behalf of the registered users. Each node is therefore associated with a set of devices/people which offers the latter the advantage of being relieved of the burden of managing their own wallet.

The accountability nodes take care of guaranteeing for the actions carried out by the IoT objects on the network; however, this is made possible by the authentication service provider which is the authority able to certify the digital identity of the device.

Typically there are different authentication service providers, and any accountability node must be able to interact with all of them. To carry out all the operations necessary to guarantee network operations, a large number of transactions must be completed, with an evident use of resources, also in terms of money. However, the required commitment is made sustainable by the fact that every time a new block is created, the profit is divided between the miner nodes and the central node, which in turn corresponds a reward to the accountability nodes that were involved during the carried-out operations. AN rewards are computed in proportion to the effort that each of them has provided in the process. This allows ANs to have enough reserves to carry out further transactions later.

The balance within the network is guaranteed by the fact that any action aimed at disturbing the stability situation within the network through actions deemed illegal is punished with the confiscation of the security deposit provided as guarantee by the AN. Furthermore, the digital identity of the owner of the AN node whose behavior was deemed incorrect is blacklisted and will no longer be allowed to generate other accountability nodes in the future.

4.1.2 Central Node

The central node exchanges information exclusively with the smart contracts that implement the network services and with the accountability nodes that it coordinates and remunerates based on the commitment made.

By virtue of its supervisory role, it certifies the security deposit paid by the ANs, after requesting accreditation and completing the registration process.

Among the various tasks to be carried out, the central node also deals with managing the smart contracts related to the execution of processes useful for providing services on the blockchain and nevertheless analyzing the flow of information exchanged with the accountability nodes, establishing which of these must be paid and which ones must instead be excluded from the network for violating the rules of conduct.

The role of central node is played by one of the ANs of the network elected according to a rotation criterion that periodically chooses the one deemed most appropriate to perform the task on the basis of a distributed voting algorithm. Finally, a further fundamental property of the central node is represented by the fact that it can be only an accountability node, which by its nature is already reliable. In addition, the application of the periodic rotation principle ensures that it does not represent a single point of system failure, which makes the entire infrastructure robust and fault tolerant.

4.1.3 Miners

The miner nodes are framed within the architecture in order to validate the transactions within ledger blocks, guaranteeing the characteristics of immutability and security. The miner is concerned with solving a complex mathematical problem to reach the consensus that leads to the validation of the block.

The miner participates in the process by making its own computing resources available to solve the mathematical algorithm that allows the validation of the blocks. Once the solution to the problem has been found, the block is transmitted to the network and this leads to the updating of the chain on all the nodes of the network, noting the information on the distributed register. The consensus mechanism described above is certainly among the most widespread, and it is known as *proof of work* (PoW); however, the miner's compensation mechanism changes according to the type of blockchain. For instance, *proof of stake* (PoS) simplifies the process described above by applying a verification process based on the distribution of the reward based on the participation fee among the individual members, which reduces the complexity of the decentralized verification process and can therefore also generate savings in energy and operating costs. Additional methods used are delegated proof of stake and proof of importance (PoI) [27].

However, a crucial aspect of the miner's job is adding the validated block to the chain. In order not to compromise the integrity of the entire network, this block must always be placed at the end of the longest already existing chain.

Since the miner's task is of absolute importance to maintain the security of the entire block chain, his effort is rewarded through a token.

The incentive that is given to miners through the token for solving the problem is the main key to ensuring that the entire system is completely reliable.

It should also be emphasized that the remuneration system changes for each blockchain and is structured around an algorithm that will have different rules that depend on the programming logic and the objectives and functions offered by the system itself.

4.1.4 Authentication Provider

Authentication providers are those entities responsible for uniquely identifying the parties involved in the cooperation process for the fulfillment of the requirements for the attribution of responsibility.

In particular, in the case of IoT devices, the goal is to verify the authenticity of PUF-embedded objects, and the role of identity provider is usually played by the producer/manufacturer, which is the only entity which is allowed to access the database storing the challenge-response pairs, obtained during the device enrollment phase previously described.

To ensure maximum security against cyberattacks that could allow the emulation of one of the authorized PUFs and therefore of the identity linked to it, this database containing the CRPs must remain confidential; therefore it is not normally transferred to third parties. The reference framework outlined leads to the consideration that each PUF manufacturer can therefore represent an independent identity provider; however, this does not preclude other settings where a manufacturer securely transfer CRPs to a third trusted party, e.g., to a consortium grouping of several producers.

As previously stated, for that regards people identification, trusted digital identities are achieved by public identity providers which operate in compliance to government regulations like the eIDAS standard [5] adopted by the European Union.

5 Conclusions

This chapter presented an infrastructure enabling accountable interactions of IoT devices participating as autonomous objects in decentralized cooperative business processes where previously interactions were only tied to legal entities like employees working for some organizations. The infrastructure exploits as the key technologies, the blockchain, and PUF-based authentication mechanisms. To maintain the level of reliability and security of these technologies, no additional software substrate were added. On the blockchain side, the infrastructure requires the introduction of ANs and of ad hoc smart contracts.

References

1. General Data Protection Regulation. Official Journal of the European Union **59**, 1–88 (2016). https://eur-lex.europa.eu/legal-content/EN/TXT/PDF/?uri=CELEX:32016R0679
2. Angiulli, F., Fassetti, F., Furfaro, A., Piccolo, A., Saccà, D.: Achieving service accountability through blockchain and digital identity. In: Lecture Notes in Business Information Processing, pp. 16–23. Springer International Publishing (2018). https://doi.org/10.1007/978-3-319-92901-9_2
3. Argento, L., Buccafurri, F., Furfaro, A., Graziano, S., Guzzo, A., Lax, G., Pasqua, F., Saccà, D.: Id-service: A blockchain-based platform to support digital-identity-aware service accountability. Applied Sciences **11**(1) (2021). https://doi.org/10.3390/app11010165
4. Bautista Adames, I.A., Das, J., Bhanja, S.: Survey of emerging technology based physical unclonable funtions. In: 2016 International Great Lakes Symposium on VLSI (GLSVLSI), pp. 317–322 (2016). https://doi.org/10.1145/2902961.2903044
5. Bender, J.: eIDAS Regulation: EID - opportunities and risks (2015). https://www.bsi.bund.de/SharedDocs/Downloads/DE/BSI/ElekAusweise/SmartCard_Workshop/Workshop_2015_Bender.pdf
6. Boneh, D., Franklin, M.: Identity-based encryption from the weil pairing. SIAM Journal on Computing **32**(3), 586–615 (2003). https://doi.org/10.1137/s0097539701398521
7. Boudguiga, A., Bouzerna, N., Granboulan, L., Olivereau, A., Quesnel, F., Roger, A., Sirdey, R.: Towards better availability and accountability for iot updates by means of a blockchain. In: 2017 IEEE European Symposium on Security and Privacy Workshops (EuroS PW), pp. 50–58 (2017). https://doi.org/10.1109/EuroSPW.2017.50
8. Buccafurri, F., Lax, G., Russo, A., Zunino, G.: Integrating digital identity and blockchain. In: OTM Confederated International Conferences" On the Move to Meaningful Internet Systems", pp. 568–585. Springer (2018)
9. Chatterjee, U., Chakraborty, R.S., Mukhopadhyay, D.: A puf-based secure communication protocol for iot. ACM Trans. Embed. Comput. Syst. **16**(3) (2017). https://doi.org/10.1145/3005715
10. Crabtree, A., Lodge, T., Colley, J., Greenhalgh, C., Glover, K., Haddadi, H., Amar, Y., Mortier, R., Li, Q., Moore, J., Wang, L., Yadav, P., Zhao, J., Brown, A., Urquhart, L., McAuley, D.: Building accountability into the internet of things: the iot databox model. Journal of Reliable Intelligent Environments **4**(1), 39–55 (2018). https://doi.org/10.1007/s40860-018-0054-5
11. Dash Core Group: Understanding masternodes. https://docs.dash.org/en/stable/masternodes/understanding.html (2021)
12. Dash.org: Dash documentation. https://docs.dash.org/en/stable/,
13. Döttling, N., Garg, S.: Identity-based encryption from the diffie-hellman assumption. In: Annual International Cryptology Conference, pp. 537–569. Springer (2017)
14. Ethereum.org: Ethereum Virtual Machine (EVM). https://ethereum.org/en/developers/docs/evm/
15. Felicetti, C., Furfaro, A., Saccà, D., Vatalaro, M., Lanuzza, M., Crupi, F.: Making IoT services accountable: A solution based on blockchain and physically unclonable functions. In: R. Montella, A. Ciaramella, G. Fortino, A. Guerrieri, A. Liotta (eds.) Internet and Distributed Computing Systems, pp. 294–305. Springer International Publishing, Cham (2019)
16. Furfaro, A., Argento, L., Parise, A., Piccolo, A.: Using virtual environments for the assessment of cybersecurity issues in iot scenarios. Simulation Modelling Practice and Theory **73**, 43–54 (2017). https://doi.org/10.1016/j.simpat.2016.09.007. Smart Cities and Internet of Things
17. Furfaro, A., Argento, L., Saccá, D., Angiulli, F., Fassetti, F.: An infrastructure for service accountability based on digital identity and blockchain 3.0. In: IEEE Conference on Computer Communications Workshops (INFOCOM WKSHPS), pp. 632–637 (2019). https://doi.org/10.1109/INFCOMW.2019.8845092
18. Gope, P., Sikdar, B.: A comparative study of design paradigms for puf-based security protocols for iot devices: Current progress, challenges, and future expectation. Computer **54**(11), 36–46 (2021). https://doi.org/10.1109/MC.2021.3067462

19. Halak, B.: Physically Unclonable Functions. Springer International Publishing (2018). https://doi.org/10.1007/978-3-319-76804-5
20. Hameed, K., Barika, M., Garg, S., Amin, M.B., Kang, B.: A taxonomy study on securing blockchain-based industrial applications: An overview, application perspectives, requirements, attacks, countermeasures, and open issues. Journal of Industrial Information Integration **26**, 100312 (2022). https://doi.org/10.1016/j.jii.2021.100312
21. Herder, C., Yu, M.D., Koushanfar, F., Devadas, S.: Physical unclonable functions and applications: A tutorial. Proceedings of the IEEE **102**(8), 1126–1141 (2014). https://doi.org/10.1109/JPROC.2014.2320516
22. IBM: What are smart contracts on blockchain? https://www.ibm.com/topics/smart-contracts
23. Idriss, T., Idriss, H., Bayoumi, M.: A puf-based paradigm for iot security. In: 2016 IEEE 3rd World Forum on Internet of Things (WF-IoT), pp. 700–705. IEEE Computer Society, Los Alamitos, CA, USA (2016). https://doi.org/10.1109/WF-IoT.2016.7845456
24. Jakobi, T., Stevens, G., Castelli, N., Ogonowski, C., Schaub, F., Vindice, N., Randall, D., Tolmie, P., Wulf, V.: Evolving needs in iot control and accountability: A longitudinal study on smart home intelligibility. Proc. ACM Interact. Mob. Wearable Ubiquitous Technol. **2**(4) (2018). https://doi.org/10.1145/3287049
25. López-Pintado, O., García-Bañuelos, L., Dumas, M., Weber, I., Ponomarev, A.: Caterpillar: A business process execution engine on the Ethereum blockchain. Software: Practice and Experience (2019). https://doi.org/10.1002/spe.2702
26. Mendling, J., Weber, I., Aalst, W.V.D., Brocke, J.V., Cabanillas, C., Daniel, F., Debois, S., Ciccio, C.D., Dumas, M., Dustdar, S., Gal, A., García-Bañuelos, L., Governatori, G., Hull, R., Rosa, M.L., Leopold, H., Leymann, F., Recker, J., Reichert, M., Reijers, H.A., Rinderle-Ma, S., Solti, A., Rosemann, M., Schulte, S., Singh, M.P., Slaats, T., Staples, M., Weber, B., Weidlich, M., Weske, M., Xu, X., Zhu, L.: Blockchains for business process management - challenges and opportunities. ACM Transactions on Management Information Systems **9**(1), 1–16 (2018). https://doi.org/10.1145/3183367
27. Mingxiao, D., Xiaofeng, M., Zhe, Z., Xiangwei, W., Qijun, C.: A review on consensus algorithm of blockchain. In: 2017 IEEE International Conference on Systems, Man, and Cybernetics (SMC), pp. 2567–2572 (2017). https://doi.org/10.1109/SMC.2017.8123011
28. Nakamoto, S.: Bitcoin: A peer-to-peer electronic cash system. https://bitcoin.org/bitcoin.pdf
29. Norta, A., Grefen, P., Narendra, N.C.: A reference architecture for managing dynamic inter-organizational business processes. Data & Knowledge Engineering **91**, 52–89 (2014). https://doi.org/10.1016/j.datak.2014.04.001
30. Ryan, P., Watson, R.: Research challenges for the Internet of Things: What role can OR play? Systems **5**(1), 24 (2017). https://doi.org/10.3390/systems5010024
31. Shamsoshoara, A., Korenda, A., Afghah, F., Zeadally, S.: A survey on physical unclonable function (puf)-based security solutions for internet of things. Computer Networks **183**, 107593 (2020). https://doi.org/10.1016/j.comnet.2020.107593
32. Singh, J., Millard, C., Reed, C., Cobbe, J., Crowcroft, J.: Accountability in the IoT: Systems, Law, and Ways Forward. Computer **51**(07), 54–65 (2018). https://doi.org/10.1109/MC.2018.3011052
33. Szabo, N.: Formalizing and securing relationships on public networks. First Monday **2**(9) (1997). https://doi.org/10.5210/fm.v2i9.548

Digital Twin Through Physical Assets Tokenization in Blockchain

Giuseppe Ferrara and Corrado Santoro

1 Introduction

One of the most relevant aspects of the brick and mortar market is that it simultaneously requires in-place presence of buyers and sellers to complete the purchase. This characteristic directly affects the business in two opposite directions. From one side, brick and mortar market allows potential buyers to directly verify the assets they would like to buy, which is a benefit that support the trustiness of the business due to the fact that buyers don't have to rely on a third party to ensure how much the item is good for them. But, on the other side, it is limited to the opening hours, its audience is strictly limited to the geolocation area, and it has high total cost of ownership *(TCO)* due to staff salary, shop rent, utility bills, and so on, which are ultimately charged on the final price tag.

At the cost of delegating the asset verification to a third part reputation, remote digital markets are instead 24/7 available with lower running costs and wider audience; thus, although it has around 20% of the market penetration [4, 39], digital markets represent an attractive alternative to the traditional ones.

The growth of e-commerce penetration in the USA, which has been continuously increasing since its appearance, has increased due to the COVID-19 pandemic by doubling in just 8 weeks as in the last 10 years, passing from 16% in 2019 to 27% in Q1 2020, and it is estimated that 95% of purchases will be facilitated by e-commerce by 2040 [11]. Since the digital market operates by exchanging coded information, it natively fits on any information-based assets such as digital licenses, tickets, or financial services. On the other hand, material assets, such as foods, buildings, or paintings within the digital market, cannot be directly manageable within any digital

G. Ferrara (✉) · C. Santoro
Dipartimento DMI, University of Catania, Catania, Italy
e-mail: giuseppe.ferrara@phd.unict.it; santoro@dmi.unict.it

L. Fotia et al. (eds.), *Security, Trust and Privacy Models, and Architectures in IoT Environments*, Internet of Things, https://doi.org/10.1007/978-3-031-21940-5_9

market. To avoid this limitation and to extend the advantages of the remote digital market also to physical assets, there are two additional steps that must be taken into account: one is the tokenization process and the other one the legitimacy of the ownership.

Tokenization is the process required to make assets available on the token digital market. It is obtained through the creation of a new digital entity called a token, which is the representation of the asset within the token digital market. In order to work properly, a token must contain the description of all the relevant features that are useful to unequivocally identify the original asset. Hence, the tokenization process is, in principle, a simple and intuitive process where IoT sensor devices can also make it automatic. In practice, however, the identification and the description of all the characteristics needed in order to identify a physical asset can actually be a very complex process which is subjected to the same challenge of the authentication problem, i.e., recognizing the right entity within a trustless environment. Although physical asset tokenization is an highly debated problem [18] due to his economic implications, according to [34] *users need to trust the correctness and accuracy of the operating algorithms, and the provision of information about the asset by trusted third parties is still an important prerequisite.*

A legitimate token is a token recognized by the market as the true single representation that detains the full ownership of the physical asset on the remote market [9, 31]. By taking its place on the digital market, a legitimate token can be traded on the digital market to transfer the ownership of the physical asset from the seller to the buyer (like the physical asset on the physical market), and this is the most relevant feature powered by this kind of token that enables the establishment of the digital twin concept for real (Fig. 1). In this scenario, for each transaction on the legitimate token, a form of guarantee must be transferred from the seller to the buyer; in this way a trusted trading transactions of legitimate tokens can take in place. The main problem here is related to the methodology/technique or algorithm used to guarantee the digital twin, i.e., the unique correspondence between physical asset and his digital representation. Indeed the token and the physical asset must be associated with a specific approach, once that the related legitimate token is on sale on the digital market. Therefore, an important question arises in this approach: how can token correspondence and legitimacy be ensured, or, in other terms, how digital twin can be established and maintained with enough trustiness to allow the trade? Therefore, an additional level of confidence is needed to allow physical assets to trade on remote digital markets.

In order to give an answer to the above question, in this work we present an architecture concept capable to enable the sales of physical assets into remote market with the lowest possible engagement of a third party. This *hybrid* architecture concept combines some promising features from different existing approaches. In particular, we first consider the state of the art of both in presence and remote market in terms of threat and opportunity, and then we analyze the current available approaches by pointing out the advantages and disadvantages.

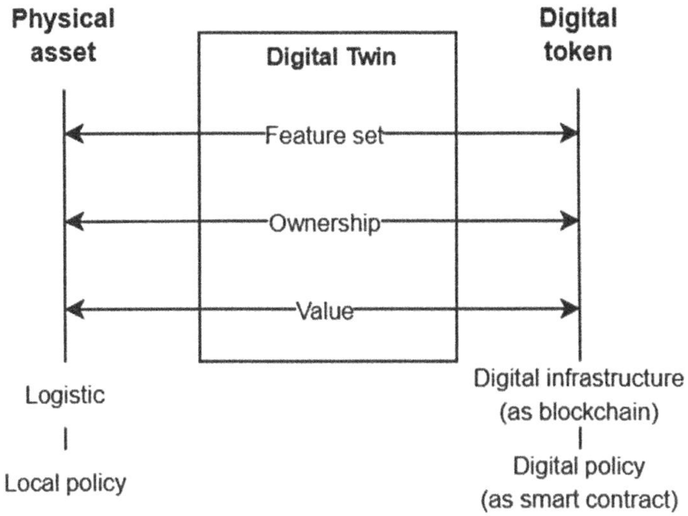

Fig. 1 Digital twin can be seen as the maintenance of the link between physical asset and digital token of features set, ownership, and value

As we discuss later in this work, the hybrid solution concept here presented can ensure physical asset binding with the digital token counterpart by leaving in-between peer-to-peer trading work, i.e., without any possible intermediary interference. The following sections are structured as follows: Sect. 2 gives an overview of the related work. Section 3 reports the state of the art in terms of available features of the existing approaches with a deep dive into the token taxonomy. Sect. 5 describes the contribution of this paper by introducing a new hybrid architecture concept that aims to represent the best trade-off between reputation and distributed reply systems. Finally in Sect. 6 we conclude by highlighting benefits and limitation of the proposed concept, as well as future works.

2 Related Work

A common approach to the problem of assets tokenization is the one based on label [20, 25]. It basically consists on the application of a label with a printed content, such as a QR code, that represents the "link" information to the token itself [13]. However, since labels can be moved, altered, or cloned, this approach is only useful to guarantee the existence of a token; it does not provide any kind of guarantee about the association of the tangible asset represented by the token, thus no real digital twin establishment.

Still based on the label approach, there are a few more sophisticated applications that use tamper-proof adhesives labels [8]. This type of labels has the characteristic to being irreparably damaged once removed, so that they cannot be removed without compromising the label itself. A more advanced application of this approach uses NFC tags or RFID labels instead of printed code to achieve better anti-forgery feature [6] that can be considered as tamper-proof labels that can solve the relabeling problem. However, the production of these labels is more complex than a simple print of a QR code and requires specific equipment that cannot be purchased and used by all potential sellers and that must rely on a third party (intermediary) for the production of these labels. Furthermore, these labels do not solve the problem of multiple labels, and, although more sophisticated, they do not even completely solve the problem of being cloned by third parties without relying on a third-party certificate authority.

Another approach to solving this problem is to rely on IoT devices to read physical features and AI as a decision engine to avoid hacks [36], but, since IoT devices struggle to provide high collateral value as a guarantee of their behavior, this approach does not work for valuable assets. Moreover, even though IoT technology can help address the problem of physical-to-digital interfacing by using IoT devices controlled by blockchain smart contracts, it does not solve the legitimacy problem itself. The Lokkit project by Siemens [36] is a perfect practical example of this, where IoT and blockchain technology combination aim to build a trust-based peer-to-peer system by excluding the trusted third party requirement. The Lokkit project defines a smart locker controlled by IoT devices and managed through smart contracts running on the Ethereum blockchain. While IoT devices are used to check the status and manage *physically* the locker through sensors, the whole back-end logic, which includes rental rules, token management, time constraints, service fee, etc., is defined through the smart contract.

Other studies [21] presented solutions in this field of research about how to quantify, map, and authenticate real or virtual assets while concluding that "trading and exchanging of Asset-backed token is still open question."

With this work we want to put the foundation stone on this open question.

3 State of the Art

Let's suppose to assign a digital token to an artwork such as a painting or a sculpture. We could describe all the physical characteristics, such as dimensions, weight, and volume, take several photographs [23], and even build an accurate 3D model [35]. Nevertheless the falsification of the original artwork is still possible, especially in case of a high valuable ones. That is why leveraging on third-party reputation as collateral value is the most reliable way, as today, to assure guarantee of authenticity, even though this system is not perfect [27]. Figure 2 shows a reputation failure example of a fake artwork that has been entrusted as original by the museum reputation for more than 2 years.

Fig. 2 The original painting (left side) is "Odalisque in Red Pants" by Henri Matisse, valuated 3 million US dollar, was stolen by swapping it with the fake painting (right side) which has been exhibited at the Contemporary Art Museum of Caracas for more than 2 years

The above example related to the artwork tokenization aims to describe the needs to put an additional level of confidence establish digital twin and enable the trading of physical assets through remote markets. A way to that is by means of a trusted third party, i.e., an intermediary. This approach is called reputation-based.

On the other hand, the *best trade* is obtained when no interference occurs between the seller and the buyer. Therefore the best scenario is known as peer-to-peer (P2P) [14, 33], including no intermediaries. A typical example of this scenario is represented by crypto currencies such as Bitcoin[26]. In the following of this section, we discuss the two different approaches, providing a brief comparison among them, in order to introduce our approach.

In order to work properly, a reputation-based system is subject to the following conditions [29]:

1. The value of the reputation used as collateral has to be higher than the value of transaction.
2. It must be always possible to publicly demonstrate any incorrect behavior of the guarantor, and, in that case, it must be possible to compromise the reputation of the guarantor.
3. The guarantor has to get enough incentive to provide the service with profit.
4. All the information required to complete the transaction is shared with the guarantor.

For convenience, we call the alternative architecture—the architecture based on blockchain [10] with no intermediaries—*distributed replay*. In this case, the involved agents sign and spread the transactions among a network where each other agent is responsible to maintain a personal copy of the same shared ledger. Once received, the transaction terms are verified locally and stored within each local copy of the shared ledger.

In order to work properly, each agent must behave in the same way so that each local copy of the ledger can be synchronized with the others. Distributed replay takes its name from the fact that all involved nodes locally replay the same operations in order to verify and store every incoming transaction.

Compared to reputation-based systems, the distributed replay architecture provides a new form of guarantee that can be trusted by all other agents/nodes without the involvement of other intermediaries. This means that a replay architecture leverages only on the *health* of the network itself (i.e., size, history, decentralization, and effort exposed to maintain it).

In order to work properly, this approach is subject to the following conditions:

1. The shared ledger is maintained by a relative majority of honest agents [5].
2. The community gets enough incentive to store and verify each transaction within the shared ledger.
3. The information required for the transaction is shared among all agents of the network.

The most noticeable benefit achieved by the tokenization of high-valued assets is the ability to fractionate ownership. This feature enabled a new business because, once an asset is tokenized and legitimated, users can easily buy even a very small portion of a high valuable asset for investment purpose and obtain eventual income in proportion, as easy as just buying a token. A new company has been founded with this specific purpose; thus this is yet available applied on diamonds, artwork, and real estate where the case of real estate is the most common.

The work of these companies is essentially about minting tokens and providing continuous proof of ownership while tokens are available. Although most of these companies leverage on peer-to-peer blockchain infrastructure to mint tokens, their users have to rely on company's reputation as collateral value of digital twin assurance, which also means trust on company's history and company's smart contract that the company uses to define the token logic.

3.1 Token

As mentioned in the introduction section, tokens are abstractions that can represent anything in both physical and digital world. This is the reason why they are valuable as long as the consensus on its abstraction is granted, so long as the blockchain can grant consensus in a trustless and permissionless environment. For this reason, tokens defined through smart contracts on top of the public blockchain are the best fitting architecture for digital tokens.

As time of writing, Ethereum is the most popular blockchain platform that supports smart contracts [12, 22, 28], so, due to the need to have a reference platform, hereinafter we will consider the Ethereum blockchain platform as a reference, but the assumptions made in this work are valid and applicable on any blockchain platform with smart contract support that can define a custom token.

We now first introduce a token classification based on their fungible ability, with focus on non-fungible tokens by going in depth on their rationale and the needs of a standard that has permitted successful applications.

3.2 *Token Classification*

A common classification used to group tokens is the one based on their fungibility attribute, i.e., the ability of a token to be interchangeable and indistinguishable from another token of the same class. To support security, reliability, and interoperability, each token class here is described by a standard, named by the progressive number of the ERC (Ethereum Request for Comments) that defines the basic structure and features of the token. In Table 1 the main Ethereum token standards are reported with sample type with respect to their fungibility classification.

3.2.1 ERC-20 Fungible Token

Fungible token, hereinafter FT, has been standardized on the Ethereum blockchain in late 2015 with ERC-20 which defines an interface that:

> Allows any tokens on Ethereum to be re-used by other applications: from wallets to decentralized exchanges [15]

Since each FT compliant with ERC-20 standard is defined with the same set of functions, platform integrations of this kind of token don't need any kind of customization. This is a successful key aspect on FT spreading that also showed how good standards can impact on spreading and how it can be considered as a use-case enabler. Figure 3 shows the amount of daily ERC-20 token transactions since the release of the standard to date. The most relevant ERC-20 applications at the time of writing are the following:

1. Initial coin offering (ICO) where fungible tokens are company equity abstractions that can be sold in change of founds. This application enabled direct P2P crowd funding without any trusted third party.

Table 1 Token classification in Ethereum

	Digital		Physical
Token classification	Asset type	Ethereum standard	Asset example
Fungible	Coins	ERC-20	Cash
Semi-fungible	Mixed tokens	ERC-1155	Person with cash
Composable	Token collection	ERC-998	Collection
NFT	Domain name	ERC-721	Artwork

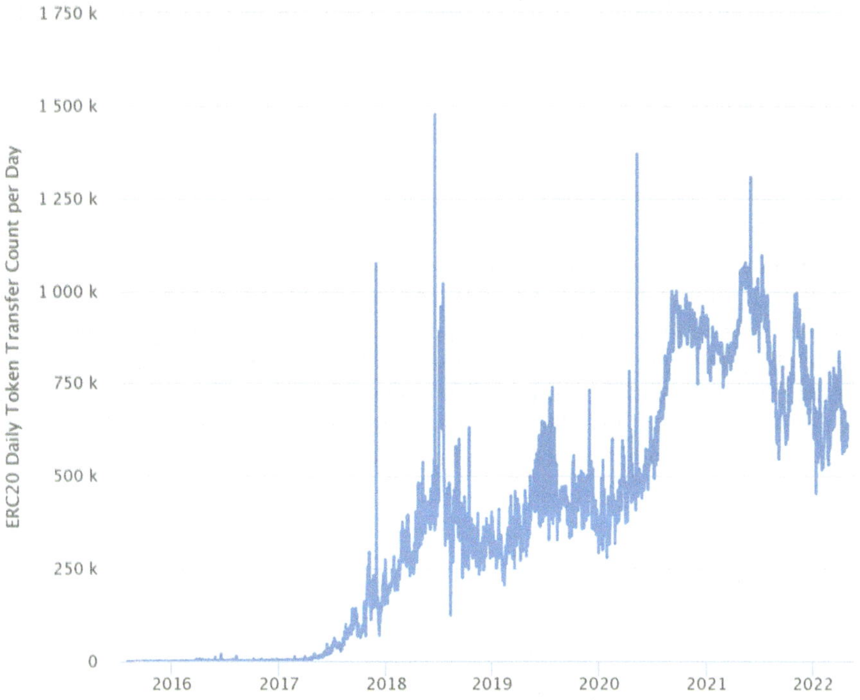

Fig. 3 ERC-20 token transaction since the release of the standard

2. Stable coin where fungible tokens are pegged on a stable asset, such as gold unit or USD, with the aim of minimize the volatility of the price, which is typical characteristic of most cryptocurrency.
3. Decentralized Autonomous Organizations (DAO) are coded organizations that run without any central governance, which is indeed managed by the token holders through a voting mechanism and where vote weight is given by token amount owned for each stakeholder.

3.2.2 ERC-721 Non-fungible Token

Non-fungible tokens, hereinafter NFTs, are, instead, an abstraction of unique items, so they can represent any item that cannot be automatically swapped 1:1 with other items. It was standardized on the Ethereum blockchain in early 2018 with the ERC-721 [37].

First known implementation of an NFT on Ethereum is within the CryptoPunk project, which dates back to 2017 [19]. The CryptoPunk token has been implemented through a smart contract as a particular type fungible token with fixed 10000 preminted units. This special fungible token has two innovative aspects that, in

Fig. 4 The original 100x100 CryptoPunk matrix. SHA256 hash of this image is hardcoded into the CryptoPunks smart contract to let the users verify the image file containing all the punks [19]

contrast with a standard fungible token, make each unit unique and distinguishable from the other 9999, so that it is not fungible anymore:

1. The hard-coded SHA256 hash corresponds to the hash of a large image representing a 100×100 matrix composed of 10.000 different 24×24px subimages in punky-look pixel art; see Fig. 4.
2. A key-value parameter named "punkIndexToAddress" that links each item of the matrix, i.e., the unique punky pixel arts, to the wallet or contract address which own it.

Similar to the ERC-20 standard for fungible tokens, the ERC-721 standard made NFT easier to use and integrate through a common interface which defines a specific functions set [30] that are mandatory to be implemented in order to be compliant.

Since unique and distinguishable items, such as artworks, places, and fingerprints, are common in both the physical and digital world, and due to the scarcity feature provided by NFT that leverages on their value, NFT received greater attention and spread by the ERC-721 standard. As available in the physical world, ERC-721 standard has enabled the marketplace capability available also on blockchain. With these marketplaces [3], all ERC-721 tokens can be sold within the same platform by enhancing their selling opportunities.

The most notable application of ERC-721 is the CryptoKitties [1] project which defines unique assets as digital kits that allow players to buy, sell, collect, and breed virtual kits as digital assets. Since NFTs are configured to act as proofs of authenticity and ownership within the blockchain network, a relevant use case to better evaluate the real impact of NFT is the name services, i.e., services like DNS where a mnemonic string is associated with a non-mnemonic parameter that, for humans, is very difficult to remember. This is the case of Ethereum Name Services (ENS) [2] where NFT owner can store a mnemonic string to resolve a wallet address which is a 40-digit HEX string.

Remarkable characteristic of services like ENS is that, unlike DNS which is censurable due to a central trusted authority management, it runs through NFTs on top of a permissionless and trustless blockchain which make it independent from any third party which also means uncensorable.

3.2.3 NFT Limitations and Opportunities

If on the one hand, the ERC-721 standard interface enables new use cases, such as the marketplace, on the other hand, this interface puts some limitations on how distinguishable tokens are defined and how they can be managed. This is the reason why other NFT standards have been proposed after ERC-721.

3.3 ERC-998 Composable NFT

This is the case of the ERC-998 standard proposal which has been defined as an extension of the ERC-721 standard in order to add the ability for NFTs to own or be owned by other NFTs and/or FTs [24]. An ERC-998 token is indeed a composable NFT that can own or be owned by other NFT, according respectively to the top-down or bottom-up approach, so that they might be also seen as a NFT collection. Since it has been implemented as an extension of ERC-721, the ERC-998 token can own the ERC-721 and ERC-20 token, but the standard ERC-721 token cannot own or be owned by the ERC-998 token.

A typical use case of composable NFT is on digital games, such as MMORPGs, where an NFT, which might refer to a special item, can be unlocked to the owner through a collection of related FTs units that might refer to coins where each unit represents a certain percentage of the item. Once the owner owns all the units of the collection, he can decide to convert them to the special item, i.e., the NFT, by burning all the FT pieces owned.

3.3.1 ERC-1155 Multi-class Token

Another remarkable example of improving existing NFT standard is the ERC-1155 [38] which provides the advantage to transfer multiple token types within a single transaction by saving gas units, i.e., total transaction cost. This is possible because token standards such as ERC-20 and ERC-721 require a separate contract for each type or collection of tokens. This gives redundant bytecode and limits certain functionality by the nature of separating each token contract into its own permissioned address.

ERC-1155 Multi-Class Token Standard allows each token ID to represent a new configurable token type which may have its own metadata, supply, and other attributes. Thanks to this characteristic, a single ERC-1155 token can include any combination of FT either NFT so that it can an represent an arbitrary number of fungible and non-fungible token at the same time. Out of transaction cost saving, the major advantages of this standard are on removing the need to "approve" individual token contracts separately and on the easier description of mixed token types within a single contract.

4 On-Chain vs Off-Chain

We now compare blockchain transactions based on distributed replay systems, also known as on-chain transactions, with transaction that rely on reputation systems, also known as off-chain transaction [17].

Once signed, a blockchain transaction is sent to all the miners to perform validation, and then it is *packed* together with all the other pending transactions to create the new last block. This process is highly redundant because there is no trusted part who indicates the right final state; thus each miner must run the same set of instructions at the same time to obtaining the same result. Although recent solutions have shown great improvements [7], this process is known to be somehow mandatory to ensure integrity within a trustless environment. This is the reason why transactions on public blockchains are expensive and needs to be optimized, as much as possible, in terms of stored data and computed data.

The main criteria used to minimize the effort required by the transaction validation on a public blockchain is mainly based on keeping only on the essential data on-chain while leaving the nonessential information *off-chain* but still linked with the essential one stored on-chain.

This idea is the basis of "Layer-2" solutions such as the Lightning Network on Bitcoin [26], or rollups on Ethereum [16]. Indeed, while on-chain transactions are slow and expensive due to the trustless environment used, off-chain transactions are faster and cheaper because they don't need to achieve distributed consensus and can use legacy central technologies such as reputation systems. On the other hand, off-chain transactions are quite unsuitable for P2P purposes due to the fact that they lack of non-repudiation assurance.

Table 2 On-chain vs Off-chain transactions

	Transaction type	
Parameters	*On-chain*	*Off-chain*
Speed	Low	High
Trust	High	Low

These different characteristics are reported in Table 2.

In order to put physical asset on sale through the digital market, the first step consists in creating his digital representation. It is therefore necessary to identify the most suitable agent for this job, that is, the one that has no advantage of interfering, although able to do so.

For obvious reasons, neither the seller nor the buyer is a good candidate as a guarantor of the one-to-one association between token and physical asset, so we can infer that at least one additional agent (or third party) must be introduced for this particular purpose.

Based on the premises above, the only form of architecture directly applicable to grant binding association through physical asset and his corresponding token is the one based on reputation and that this requires the involvement of an auxiliary trusted intermediary. The reputation system is the best compromise to achieve our goal and, at the same time, limit as much as possible the potential interference of the intermediary.

In the next section, we deal with this aspect by providing a hybrid solution that represents, in our view, a good compromise between distributed replay and reputation-based systems to solve the problem we have set above.

5 Concept Architecture to Establish Digital Twin on Physical Asset Tokenization

The solution we present in this section has been designed to satisfy the main requirements discussed in the previous section:

1. Limiting, as much as possible, the interference of any third party.
2. Guarantee the minimum level of confidence on asset-to-token binding.

To achieve these points, we have performed an analysis of the tokenization process for a physical asset, and, after that, we have designed a specific solution organized into three phases: phase 1 and phase 3, which represent the beginning and the end of the process itself, are designed to be reputation-based, while phase 2 leverages on distributed replay.

As we discuss later in this section, in our approach the intermediary's domain is limited to those activities related to the transaction, acting as a provider of a fiduciary and logistic service. For the remainder of this section, we denote, for convenience, as WT (*warehouse tokenizator*) the intermediary of the transaction whose role is

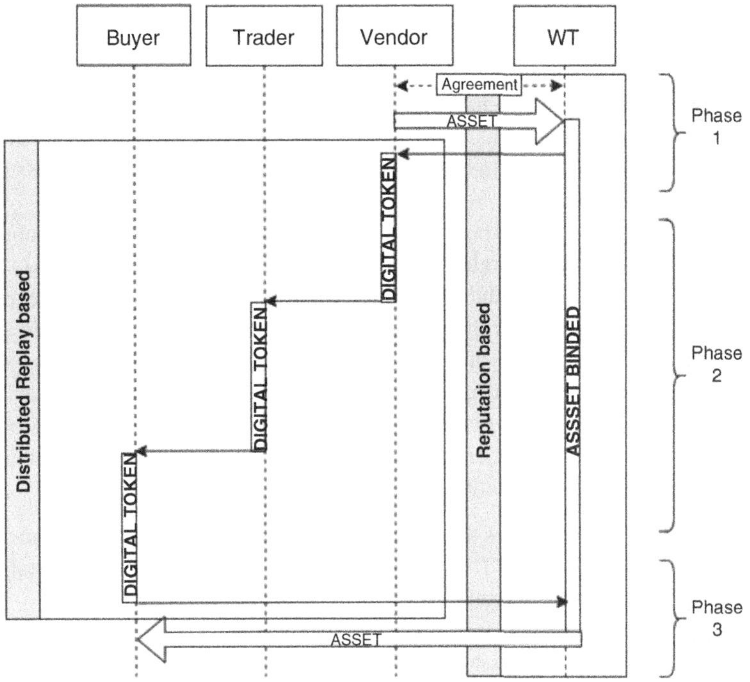

Fig. 5 The three phases of the hybrid solution with the application domains

to do business by providing token legitimacy as the exclusive representative of the ownership of the physical asset.

In our approach, WT will be responsible for safeguarding the physical asset until the token is back to him (after that the asset has been sold) or the agreed terms have expired. In this way, as we discuss later in this section, the token can work as a proof of asset availability.

We also observe that, unlike most digital assets, such as cryptocurrencies, each physical asset is unique, i.e., non-fungible [32]. For this reason, the tokenization process of a physical asset must also produce a non-fungible token such as ERC-721 or ERC-1155.

The proposed solution schema is depicted in Fig. 5 and detailed in Sects. 5.1, 5.2, and 5.3.

5.1 Phase 1

In the first phase (Fig. 5), the vendor agrees a set of terms of the service, and, thereafter, it moves the asset into the WT location. Once the asset has been received, the WT performs a check of all its characteristics and, if they are compliant with the agreed terms, it mints a unique digital token. The ownership of the token is set to the identity of the vendor. Now, the vendor can use the new digital token in place of the asset.

The key aspect, in this phase, is represented by the fact that the vendor trusts the WT, which acts as "safeguard" of the asset in its own warehouse. In this scenario, the vendor trusts the WT due to the reputation of the WT, which, in turn, assumes a crucial role in the transaction.

Since the main interest of the WT is to make business and, in turn, to not incur in loss of reputation, it is highly desirable that the WT reputation has a value comparable with the asset value. This mechanism should guarantee to legitimate the digital token as exclusive representative of the ownership of the tangible asset.

Moreover, by moving the physical asset into the WT warehouse, the vendor agrees with the WT a set of terms for asset storage and tokenization. Anyway, while the physical movement of deposit is achieved off-chain for obvious reasons, the agreement can be made in either ways on-chain or off-chain as long as vendor can publicly leverage on the reputation of the WT.

Feedback This phase includes the exchange of two different feedbacks. The former is left by the vendor to the WT once the vendor itself had received a digital token that is compliant with the agreed terms.

This first feedback will be autogenerated, because the compliance of the digital token can be checked automatically. If there is a complete match between the agreement terms and token characteristics, a positive feedback will be released. In this manner, the vendor cannot act in a dishonest manner by leaving a negative feedback to spoil the reputation of the WT.

Finally the vendor must pay the WT for the service (the digital token), and, as a consequence, the WT will (automatically) leave a (positive) feedback once the payment has been performed. Moreover, in order to keep the record of the feedback history and apply these rules with transparency, this solution is applicable only in case the agreement is on-chain, i.e., managed by a smart contract on a blockchain that leverage on distributed replay system.

5.2 Phase 2

In the second phase, vendor is legitimated to trade the digital token obtained during the previous phase in the digital market by leveraging on WT reputation collateral. In this case, the current token owner can simply leverage on the distributed replay system. By exclusive use of distributed replay assurance, any other intermediary is completely excluded because, since the digital twin is independently assured in phase 1, there is no need to involve them.

By phase 2, we achieve our ultimate goal of enabling P2P trading of physical assets by legitimately transfer ownership without intermediaries. The last pending point is to manage the switchover by providing the physical asset in change of the legitimate digital token, i.e., the next and last phase.

5.3 *Phase 3*

In phase 3 the buyer, who has acquired the token on the digital market, must exchange the token with the its corresponding physical asset (Fig. 5). In order to achieve this final task, since the buyer was not known at the time of the tokenization agreement, the transfer of the physical asset to the buyer location must be organized and performed by the buyer itself.

Once token ownership is set to his minter, buyer is automatically qualified to claim the physical asset and to leave a feedback to WT about the token received and feed his reputation. Moreover, buyer can always give a proof of being the last owner because the entire history of the token is publicly available and verifiable on the reference blockchain that is an append-only data structure. Furthermore, once the token ownership returned to the minter (the WT), the binding between the token and the asset is lost, and the token will be automatically destroyed.

Feedback In this phase, two other feedbacks are exchanged. The former is released by the WT to the buyer once the token has been sent back to him by the buyer itself. Also this (positive) feedback is automatically generated. The second feedback is released by the buyer to the WT once the buyer itself has acquired the asset itself.

Since the asset transfer process is performed off-chain, this feedback cannot be generated automatically. Therefore, the buyer must manually send the feedback to the WT. In this case, in order to avoid malicious buyers' behaviors aimed at spoiling the reputation of the WT with negative feedback associated with low-value transactions, the feedback must be weighted with the value of the asset and the reputation of the buyer itself.

5.4 *Limitations*

Although the proposed solution, as described above, is potentially applicable on any type of physical assets, it however comes with some intrinsic limitations that affect the final usefulness of the proposed solution.

The first one is related to the potential physical limitations of the asset itself, by way of example, tokenized physical assets can be perishable in an unpredictable way or being wrongly evaluated similarly to what happened in the painting example reported in Sect. 3 Fig. 2 so, even if the token owner is guaranteed by WT reputation, the buyer always has to consider these risks before proceeding with the purchase.

The second limitation is logistic. As discussed, during the whole life of the token, the physical asset is guarded by WT within his warehouse, but, once it is claimed in phase 3, buyer becomes the new owner; hence he must manage the asset transport to his desired location or agree with WT new terms for custodial. Since location information, like GIS position, can easily be natively embedded within the digital token (because it is a precise information), also this second limitation is known a prior, and it can be managed in advance of the trade.

The third is the local regulation. Since assets on digital twin exist on both digital and physical dimensions at the same time, they are also subject to both governance regulation and restrictions simultaneously. In fact, vendor could potentially ask to tokenize illegal or stolen assets, but, since WT role has been defined to avoid this circumstance, this scenario should not happen due to WT reputation value at risk of negative feedback. Therefore, even though this limitation is managed, it is not 100% guaranteed that this case cannot occur.

The last and most important limitation is related to the initial WT network. As well as any other business, the WT business works fine once it is fully operational but hard at early stage, i.e., when no or few WT services are available and when no or low reputation value is available as collateral. Although this is not a technical limit and out of the scope of this work, we see this point as the biggest threat to the practical use of this solution.

6 Conclusions

In this work we have discussed the state-of-the-art about physical asset tokenization by pointing out the relevant aspects by discussing benefits and drawbacks of some relevant available solutions also based on IoT technology. It has also been presented a concept architecture to enable peer-to-peer physical asset trading through the establishment of legitimate digital twin. The proposed solution is based on the combined usage of a reputation system and the distributed replay, thus also off-chain and on-chain transactions. By leveraging on reputation-based mechanisms, the architecture ensures that the digital representation is always bonded to the physical asset, i.e., it is legitimate, while, indeed, leveraging distributed replay keeps the digital representation valid for peer-to-peer trading on blockchain.

We have shown how a reputation system part is mandatory for physical asset tokenization and how it can be coupled with P2P systems to enable asset trading without third-party interference. We have discussed the details of this concept architecture and how it can avoid some of the issues that afflict monolithic solutions. We also discussed the most relevant risks related to the use of this system and described how and which of these risks and limitations can be mitigated through the proposed approach and which not. In future work, our aim is to define the architecture in low-level detail to implement a prototype of this concept and to spot the most relevant applications that can take advantage of the proposed solution. To achieve this goal, the next step is to work on a specific case study so that the approach can be experimentally evaluated.

References

1. Crypto kitties technical details, 2017.
2. Ethereum name service (ens) a distributed naming system based on the ethereum, 2017.
3. Opensea, the world's first and largest digital marketplace for crypto collectibles and non-fungible tokens (nfts), 2017.
4. GA Agency. The growth of ecommerce sales in the us.
5. Teresa Alsinet, Josep Puyol-Gruart, and Carme Torras. *Artificial intelligence research and development: proceedings of the 11th International Conference of the Catalan Association for Artificial Intelligence*, volume 184. IOs Press, 2008.
6. Naif Alzahrani and Nirupama Bulusu. Securing pharmaceutical and high-value products against tag reapplication attacks using nfc tags. In *2016 IEEE International Conference on Smart Computing (SMARTCOMP)*, pages 1–6. IEEE, 2016.
7. Eli Ben-Sasson, Iddo Bentov, Yinon Horesh, and Michael Riabzev. Scalable, transparent, and post-quantum secure computational integrity. *IACR Cryptol. ePrint Arch.*, 2018:46, 2018.
8. David Olivier; Jaquet Chiffelle; Eoghan Casey; Jonathan Bourquenoud. Tamperproof timestamped provenance ledger using blockchain technology. *Forensic Science International: Digital Investigation*, 33:300977, 2020.
9. Vitalik Buterin. The most important scarce resource is legitimacy, Mar 2021.
10. Christian Cachin and Marko Vukolić. Blockchain consensus protocols in the wild, 2017.
11. Daniela Coppola. E-commerce share of uk retail sales 2019-2024, Jul 2021.
12. Crypto Crunch. The top five smart contract platforms, 2021.
13. Thomasson E. Carrefour says blockchain tracking boosting sales of some products, 2019.
14. Liran Einav, Chiara Farronato, and Jonathan Levin. Peer-to-peer markets. *Annual Review of Economics*, 8(1):615–635, 2016.
15. Vitalik Buterin Fabian Vogelsteller. Erc-20 token standard, 2017. See also URL https://eips.ethereum.org/EIPS/eip-20+.
16. Ethereum Foundation. Ethereum l2 scaling rollups, 2021.
17. Lewis Gudgeon, Pedro Moreno-Sanchez, Stefanie Roos, Patrick McCorry, and Arthur Gervais. Sok: Off the chain transactions. *IACR Cryptol. ePrint Arch.*, 2019:360, 2019.
18. O. Ross; J. Jensen. Assets under tokenization: Can blockchain technology improve post-trade processing? In *ICIS 2019 Proceedings*, 2019.
19. Matt Hall John Watkinson. Cryptopunksmarket.sol line 5, 2017. See also URL github.com/larvalabs/cryptopunks/blob/master/contracts/CryptoPunksMarket.sol#L5+.
20. Mark Kim, Brian Hilton, Zach Burks, and Jordan Reyes. Integrating blockchain, smart contract-tokens, and iot to design a food traceability solution. In *2018 IEEE 9th Annual Information Technology, Electronics and Mobile Communication Conference (IEMCON)*, pages 335–340. IEEE, 2018.
21. Xuefeng Li, Xiaochuan Wu, Xin Pei, and Zhuojun Yao. Tokenization: Open asset protocol on blockchain. In *2019 IEEE 2nd International Conference on Information and Computer Technologies (ICICT)*, pages 204–209. IEEE, 2019.
22. M.Bartoletti; L.Pompianu. An empirical analysis of smart contracts: Platforms, applications, and design patterns. 2017.
23. Magnus: Shazam for art.
24. Jordan Schalm Matt Lockyer, Nick Mudge. Eip-998: Erc-998 composable non-fungible token standard, 2018. See also URL https://eips.ethereum.org/EIPS/eip-998.
25. Ahmed S.; Broek N. Blockchain could boost food security. *Nature*, 550(43), 2017.
26. Satoshi Nakamoto. Bitcoin: A peer-to-peer electronic cash system, 2009.
27. William Neuman. Topless woman found. details sketchy.
28. Emma Newbery. 6 top cryptocurrencies with smart contracts, 2021.
29. Beng Chin Ooi, Chu Yee Liau, and Kian-Lee Tan. Managing trust in peer-to-peer systems using reputation-based techniques. In *International Conference on Web-Age Information Management*, pages 2–12. Springer, 2003.

30. OpenZeppelin. Required interface of an erc721 compliant contract.
31. Elizabeth G. Pontikes. Two sides of the same coin: How ambiguous classification affects multiple audiences' evaluations. *Administrative Science Quarterly*, 57(1):81–118, 2012.
32. Ferdinand Regner, Nils Urbach, and André Schweizer. Nfts in practice–non-fungible tokens as core component of a blockchain-based event ticketing application, 2019.
33. Detlef Schoder and Kai Fischbach. Peer-to-peer prospects. *Commun. ACM*, 46(2):27–29, February 2003.
34. Notheisen; Cholewa; Shanmugam. Trading real-world assets on blockchain. *Business & Information Systems Engineering*, 59(6):425–440, 2017.
35. Filippo Stanco, Sebastiano Battiato, and Giovanni Gallo. *Digital imaging for cultural heritage preservation: Analysis, restoration, and reconstruction of ancient artworks*. CRC Press, 2017.
36. Tim Weingärtner. Tokenization of physical assets and the impact of iot and ai. In *European Union Blockchain Observatory and Forum*, volume 10, pages 1–16, 2019.
37. W.Entriken;D.Shirley;J.Evans;N.Sachs. Eip-721: Non-fungible token standard, 2018. See also URL https://eips.ethereum.org/EIPS/eip-721.
38. W.Radomsk;A.Cooke;P.Castonguay;J.Therien;E.Binet;R.Sandford. Erc-1155 multi token standard, ethereum improvement proposals, 2018. See also URL https://eips.ethereum.org/EIPS/eip-1155.
39. Kyna Ysabel. E-commerce vs. brick & mortar: The state of retail in 2021.

Security in Home Automation

Julian Calleja, Lalit Garg, and Vijay Prakash

1 Introduction

Home automation is not new and has been around since the 1970s. Due to technological advancements, users' perceptions of what the home should do and how services are provided have changed through the years [15]. Technology has made its way in incremental steps due to the early challenges encountered by these home automation systems (HAS). Challenges included high manufacturing and development costs[13, 14]. Due to these challenges, early home automation systems were costly, which meant that it was not attainable by the average consumer. However, prices have decreased recently, and demand increased, increasing devices introduced to the automation market. A study showed that nearly 20% of homeowners use at least one automated feature in their home and about 60% are interested in having a connected home. With these adoption rates and advancements, the market is projected to reach $91,645 million by 2023, growing at a compound annual growth rate of 11.2% from 2017 to 2023. Therefore, these figures show that HAS are essential to users' lives [1]. According to the previous study, the top factors impacting the home automation market include an increase in the importance of monitoring the home from remote locations and a rise in awareness related to the availability of these devices, such as automated security and lighting systems. These are all factors for the HAS market growth with the help of the Internet of Things (IoT). Therefore, this suggests that such systems' adoption is increasing since the users want to benefit from such features that allow them to control their home

J. Calleja · L. Garg (✉)
Faculty of Information and Communication Technology, University of Malta, Msida, Malta
e-mail: lalit.garg@um.edu.mt

V. Prakash
School of Computer Science, University of Petroleum and Energy Studies (UPES), Dehradun, India

L. Fotia et al. (eds.), *Security, Trust and Privacy Models, and Architectures in IoT Environments*, Internet of Things, https://doi.org/10.1007/978-3-031-21940-5_10

appliances from a single place in their daily lives. However, as the adaptation of these systems is increasing, security issues are also becoming a concern. In today's world, hackers do not need to travel to target homes and can easily target homes through virtual attacks. Examples can be seen in recent phenomena like when 50,000 printers were hacked, allowing the hacker to print a message. The same hacker also targeted several smart televisions to display a message. Thus these issues show that security is a significant factor that users consider when adopting a smart home.

2 Technology and Features

The introduction of the IoT helped revolutionize a new era in technology for the HAS. The technology could allow users to develop systems that were considered fiction into reality through developments of wireless interconnected, automated systems that can be managed using mobile devices. IoT presents a variety of technologies available in the market that can aid in developing a system that enables users to control their homes. Some of the available devices that the user can make use of can be categorized in the following sections:

2.1 Off the Shelf

Devices such as Samsung's SmartThings [25], Amazon's Alexa [8], Google Home [20], and Apple HomeKit are all examples of off-the-shelf devices used in a standalone form such as Google Home and Alexa or integrated with other devices to provide a better user experience. Integration between well-known devices can sometimes be considered an easy process because most manufacturers will provide integration with other devices; therefore, users will usually opt for off-the-shelf devices when starting.

2.2 Do-It-Yourself (DIY)

This is another route that the user can take. If the user adopts this approach, the user has to buy sensor devices for temperature, light, infrared, etc. and make these sensors communicate and integrate to build a HAS. This route is more intensive and often time-consuming because the user has to make these different sensors communicate with each other. Thus if the user decides to go down this path, then the use of software and hardware that allows devices to communicate with each other must be made use of. A benefit the user gains when using such a system is that it enables more flexibility, thus allowing the user to customize the system as desired.

On the other hand, this can be more challenging to set up and maintain; therefore, it requires the user to be technology-oriented [3, 30].

2.3 Hybrid

This is when a system comprises both off-the-self devices and device sensors. A system that can serve as a bridge between off-the-shelf and other devices that can be used is Home Assistant (HA). HA is an example of open-source software that allows the user to have a platform where different devices can communicate while also allowing the user to write automation. Thus, this system can serve as a bridge between off-the-shelf devices and other devices not supported by the off-the-shelf, thus helping bridge the communication between mentioned devices [19].

Features highly depend on the available devices and how well these devices integrate. Another factor that affects the features is the approach the user has adopted. With the off-the-shelf route approach, certain features might not be available due to the lack of integration between different devices. In contrast, the DIY or hybrid approach can allow the user to be less restricted. In recent years, wireless home networking technologies such as remote control have grown in popularity. Similarly, the incorporation of wireless technology in automation systems provides a number of benefits that could not be attained with a wired network alone [14]:

1. Reduced installation costs: There is no need for cabling; therefore installation expenses are drastically reduced.
2. Scalable and expandable: The Compare of Wireless network is extremely important when a network expansion is required due to new or altered requirements.
3. Internet connectivity: Control smart home equipment from anywhere in the globe with a mobile phone.
4. Security: Easily connect devices to form an integrated smart home security system, and the smart home's inherent security maintains its integrity.

3 Advantages and Limitations

Home automation offers a variety of benefits to a homeowner. Some of these features allow the homeowner to keep track of different aspects of the home. These can all be done remotely, thus allowing the user to monitor the premise, switch off the lights, control appliances, and even lock and unlock doors. Other advantages of the inhabitant's lifestyle are improved through the following examples: savings, convenience, control, comfort, and peace of mind. All of these mentioned advantages provide the user with a better living experience.

In this area, one can find various examples when working with HAS. These examples can all vary and be used interchangeably, allowing the user to build a system suitable to the specific needs rather than an already developed system. The user does not have any freedom regarding how it is implemented. One must also consider that even though the different users can have the same devices, the way users use these devices will be different [4].

HAS can face some challenges. Two main challenges which are faced are privacy and difficulty in achieving security. Security is a very challenging area because attacks are continuously being deployed and thus pose a threat to these systems, which can cause harm to the actual users that are using the system.

Another issue crucial for users is privacy. Hackers can exploit user privacy to their advantage if a smart home is not fully protected. This is all caused because we are becoming more connected and thus providing more ways of stealing our data. Security must be taken very seriously when implementing a smart home [30].

4 Literature Review

Various contributions in technology helped have to gain popularity. These contributions include increased processing power found in the processors used by these devices. Although there is an increase in processing power, there was a considerable reduction in the power consumption, cost, and size of these new electronics [15]. Also, the use of wireless technology incorporated into these sensors helped incorporate them in every aspect of the home, thus allowing the users to understand better and control their home.

Some aspects that these devices allow the user to control are the ability to know if a door or window is open or if the lights are on. These devices have more processing power to allow for a considerable amount of data being available to the users to keep an eye on their home through live feeds of video and audio and also through the use of dashboards with the help of other devices such as microphones, cameras, and other device sensors. With all this data available and stored in a database, the homeowner can access this data from anywhere. However, one cannot forget that if the user from these remote locations can access the device, it can also provide the possibility for the devices to be accessed by other users, such as hackers.

The data can contain a variety of information regarding the residents of the home that can expose the daily routines and habits of the home inhabitants [15]. All of this data generated can bring many different security vulnerabilities to the HAS. Thus, if this data is compromised, this could lead to a more significant security threat to the user's privacy and physical well-being. The advantage of this system over similar current systems is that the user can receive warnings and status updates delivered by the Wi-Fi-connected microcontroller-managed system on his phone from any distance, regardless of whether his phone is connected to the Internet.

A smart house is a residence outfitted with intelligent technologies designed to provide personalized services for its occupants. Smart technologies allow for the monitoring, control, and assistance of residents, which can improve the quality of life and encourage independent living. It is necessary to study the user's perspective and the existing state of smart homes in order to assist the installation and adoption of smart home technologies. Given the rapid development of the literature in this field, there is an urgent need to review the literature [18].

Internet of Things (IoT) refers to the concept of remotely connecting and monitoring physical items (things) via the Internet. This approach can be effectively applied to our home to make it smarter, safer, and more automated. This IoT project focuses on developing a smart wireless home security system that provides alerts to the owner via the Internet in the event of a break-in and can optionally sound an alarm. Additionally, the same collection of sensors can be employed for home automation [17]. The Internet of Things is a system in which software, sensors, and actuators are embedded within appliances. The gadgets can communicate with one another and transfer data through a network. This technology is implemented in our home to automate and facilitate the operation of appliances. The authors focus on constructing a wireless home security system. Utilizing AES encryption, network security is accomplished. In the event of a break-in, Internet-based alerts are sent to the homeowner, and an alarm can be activated if necessary. Utilization of home automation is enabled by installing appropriate sensors around the home [27].

Due to this data being available, the security concept also had to change. Security changed from the simple lock and key security concept to adopting a security system that uses cameras, microphones, proximity sensors, alarms, and silent alarms. Vulnerabilities are continuously being exploited, and attacks against flaws in such devices are constantly being deployed. These attacks can vary from attacking the software that controls the devices to the actual physical hardware. Researchers observed attacks on the software that controls these systems. There is a high probability that it is built using web technologies as a result of these devices using web technologies as their communication medium. These researchers noticed that communication between the user and the system is done over HTTP rather than HTTPS. It allows an attacker to use an array of attacks, including eavesdropping on the communication and gathering the user information, such as log-in credentials. Other systems will enable the user to authenticate through an authentication cookie without any constraints such as session or time; therefore, if a hacker manages to get hold of the cookie, then he has access to the system bypassing every authentication [7, 24]. At the same time, other research outlined how Cross-Site Scripting (XSS) and buffer overflow attacks can be exploited to get information regarding the HAS.

Attacks targeting the communication network between the devices were also explored. The research discovered the possibility of two types of class attacks against sensor networks that were not previously documented: sinkhole attacks and HELLO flood attacks. The main aim of sinkhole attacks is to get hold of the traffic from a specific location by using an affected node in the HAS, while HELLO flood attacks are possible because protocols depend on nodes to broadcast HELLO packets to ensure that neighbors receiving these packets can assume that they are in the normal range of the sender. This can be an issue because a hacker can persuade the nodes found in a network to be vulnerable to this attack and thus get hold of the users' data. These attacks mainly affect sensor networks because of their particular communication pattern. This research also discussed other attacks, such as selective forwarding, spoofed, Sybil attacks, wormholes, and acknowledgement spoofing [16]. Research on how the ZigBee wireless network protocol can be hacked using replay attacks was also examined. This attack takes advantage of the reflashing of new keys that were being transmitted in plain text over the air. A hacker could take advantage of the vulnerability, get hold of encryption keys, and do various activities such as injecting, decoding, and altering data packets to change how these devices act. Another attack possible on the ZigBee wireless protocol is eavesdropping on encrypted or encrypted traffic [21].

Another research carried out by IOActive, Inc. discovered several vulnerabilities in home automation devices developed by Belkin WeMo. Belkin is a company that provides home automation devices such as power outlets and light switches that allow the user to access these devices through an application remotely. The issues that were found could affect up to 500,000 users, providing the attackers with the capability to gain remote access to these devices connected through the Internet, install malicious firmware updates, and even access the home network. Thus, these issues show how security can be an issue when working with a HAS.

In recent years, home automation systems have garnered significant scientific interest. It aids us in living a comfortable existence and progressively improves our lifestyle quality. The many approaches utilized by this system have been analyzed. In recent times, a smartphone with an Android application is used to monitor and operate the home automation system's appliances [26]. One cannot forget how the possibility of simple devices already found in the household linked to the Internet with the help of IoT can affect these household devices so that they can be exploited to cause physical damage to the inhabitants. An example of this was demonstrated in research that explored how using smart connected lights can be dangerous to users who have epilepsy [22].

Mobile phones can also threaten a HAS because mobile is always with the user, thus connecting to different networks. A hacker trying to hack a user's HAS can use these devices as a gateway. This is a significant issue that users have to consider because some users tend to care less about what they do with their phones than they do about their homes. As we can see, an incredible amount of research has been done toward developing and monitoring attacks targeted at HAS. Still, due to the rapid advancements in technology, security must be kept in mind so that the threats mentioned and others continuously being deployed can be avoided, thus providing the user with better security and privacy. This author discusses electrical home automation equipment. Specifically, it describes several household sensors that can transform a conventional home into a "smart home." In addition, it discusses their potential applications, such as ventilation, security, and temperature monitoring. For each sensor, the article offers low-cost sensory devices, such as an Arduino and a Raspberry Pi, to set up inexpensive home automation projects [11].

The majority of Internet of Things home automation wireless technologies are based on ZigBee, Z-Wave, Wi-Fi, and Bluetooth wireless technologies. Even though these solutions are adequate, consumers of smart homes face difficulties in picking the optimal technology. This research analyzes the user-perceived performance of ZigBee, Z-Wave Wi-Fi, and Bluetooth technologies. As guidelines for possible users, six indices were created: power consumption, range, cost, ease of use, scalability, and interoperability [6].

The authors [28] proposed a smart, energy-efficient home automation system that can be accessed and controlled from anywhere in the world. For this system, an Internet-accessible Internet connectivity module is coupled to the main supply unit of the home system. A static IP address is employed for wireless communication. Home automation is based on a multimodal application that may be controlled by the user's voice using the Google Assistant or through a web-based application. Some of the home automation technology has been tabulated in Table 1.

Kitchen Genie [10] is an intelligent kitchen that proposes a technology that would intelligently transform the household, particularly the kitchen. This will be accomplished by automatically tracking inventories without requiring the user to interact with the products. This technology would not only be beneficial in the home, but it could also be developed for use in commercial kitchens or supermarkets. It might be especially useful for families in which both parents work and, upon returning home, must complete all housekeeping in addition to grocery shopping, etc. Consequently, Kitchen Genie would save a great deal of time. Those with a huge storage area could save a substantial amount of money as well. This makes it difficult to maintain track of their inventory, and they wind up discarding things that have passed their expiration date. This system utilizes a NEX-6810 Ultra-High Frequency Mid-Range Radio Frequency Identification Reader and Higgs3.

5 Conclusions

Security in the home automation field must be taken seriously by both the manufacturers and the users. Manufacturers have to make sure that they test out their systems and continuously update their techniques when new vulnerabilities are found. This ensures that the users can be more okay with incorporating these systems into their daily lives, reducing threats that the user might face.

Another threat that the manufacturers and the users have to keep in mind is that if the hacker compromises a HAS, he can cause a wide range of damage. It can vary from physical harm, emotional trauma to the home's inhabitants, and permanent damage to both devices and the premise. Therefore, these threats should not be taken lightly considering such a system.

From the users' perspective, one must ensure that devices are updated with the latest software and that passwords are changed. Therefore, this means that users have to become more educated about implementing these systems and better understand what to do to monitor their systems against these threats.

Finally, manufacturers must keep in mind that these systems will be used by people with varying levels of technical knowledge. Therefore, manufacturers must consider that not everyone will be as careful about security.

Table 1 Comparison of various home automation system technologies and controlling devices

Authors	Technology	Controlling devices	Merits	Demerits	Applications
Adhiya et al. [29]	Wi-Fi and Arduino	Web application	Cheap and secure	Sent IoT data can be hacked, resulting in security or privacy breaches. Errors are increasingly probable as system complexity increases	Monitor and control the appliances based on temperature and motion detection
Muhammad & Khalil [2]	Bluetooth and arduino	Mobile with pythons	Secure and less costly	Low bandwidth and very less communication range restricted to 10m with less data rate	Controlling the appliances
Farhan [23]	Interface card, web server,and Rasberry pi	Android app	Autonomous and scalable	Less secure due to wireless connection	Controlling the device interface
Baraka et al. [3]	Arduino and wireless ZigBee and wired X10	Android app	Scalable and energy-efficient	High replacement costs due to Zigbee and X10 modules can be used for error-tolerant systems	Scheduling of appliances using less cost
Cui et al. [5]	Hadoop and cloud-based server	Smart device	Efficiently manages unstructured and semi-structured data	Need high-speed and reliable Internet connection. Vulnerable to attacks	Monitor and control the home appliances
Gabriele et al. [9]	Cloud-based server and PCB circuit	Android app	Energy efficient	Very high cost of components and placement of the different sensors is very crucial to get the maximum coverage	Monitor and control the home appliances and respective power consumption
Gebhardt et al. [12]	Konnex bus and Raspberry Pi	Mobile Application	Less power consumption	Costly than Arduino-based systems	Monitor and control the temperature and lighting system

References

1. Home automation market size and forecast, accessed May 21, 2022.
2. Muhammad Asadullah and Khalil Ullah. Smart home automation system using bluetooth technology. In *2017 International Conference on Innovations in Electrical Engineering and Computational Technologies (ICIEECT)*, pages 1–6. IEEE, 2017.
3. Kim Baraka, Marc Ghobril, Sami Malek, Rouwaida Kanj, and Ayman Kayssi. Low cost arduino/android-based energy-efficient home automation system with smart task scheduling. In *2013 Fifth international conference on computational intelligence, communication systems and networks*, pages 296–301. IEEE, 2013.
4. Pedro Chahuara, François Portet, and Michel Vacher. Context-aware decision making under uncertainty for voice-based control of smart home. *Expert Systems with Applications*, 75:63–79, 2017.
5. Yun Cui, Myoungjin Kim, Yi Gu, Jong-jin Jung, and Hanku Lee. Home appliance management system for monitoring digitized devices using cloud computing technology in ubiquitous sensor network environment. *International Journal of Distributed Sensor Networks*, 10(2):174097, 2014.
6. Salim Jibrin Danbatta and Asaf Varol. Comparison of zigbee, z-wave, wi-fi, and bluetooth wireless technologies used in home automation. In *2019 7th International Symposium on Digital Forensics and Security (ISDFS)*, pages 1–5. IEEE, 2019.
7. Jose Fonseca, Marco Vieira, and Henrique Madeira. Testing and comparing web vulnerability scanning tools for sql injection and xss attacks. In *13th Pacific Rim international symposium on dependable computing (PRDC 2007)*, pages 365–372. IEEE, 2007.
8. Marcia Ford and William Palmer. Alexa, are you listening to me? an analysis of alexa voice service network traffic. *Personal and ubiquitous computing*, 23(1):67–79, 2019.
9. T Gabriele, L Pantoli, V Stornelli, D Chiulli, and M Muttillo. Smart power management system for home appliances and wellness based on wireless sensors network and mobile technology. In *2015 XVIII AISEM annual conference*, pages 1–4. IEEE, 2015.
10. Lalit Garg, K Ramesh, Gaurav Garg, Allene Portelli, and Arshi Jamal. Kitchen genie: An intelligent internet of things system for household inventory management. In *Proceedings of ICETIT 2019*, pages 3–20. Springer, 2020.
11. Alexandros Gazis and Eleftheria Katsiri. Smart home iot sensors: principles and applications- a review of low-cost and low-power solutions. *International Journal on Engineering Technologies and Informatics*, 2(1):19–23, 2021.
12. Jan Gebhardt, Michael Massoth, Stefan Weber, and Torsten Wiens. Ubiquitous smart home controlling raspberry embedded system. In *UBICOMM: The Eighth International Conference on Mobile Ubiquitous Computing, Systems, Services and Technologies*, 2014.
13. John J Greichen. Value based home automation for todays' market. *IEEE Transactions on Consumer Electronics*, 38(3):XXXIV–XXXVIII, 1992.
14. Vaishnavi S Gunge and Pratibha S Yalagi. Smart home automation: a literature review. *International Journal of Computer Applications*, 975(8887–8891), 2016.
15. Arun Cyril Jose and Reza Malekian. Smart home automation security: a literature review. *SmartCR*, 5(4):269–285, 2015.
16. Chris Karlof and David Wagner. Secure routing in wireless sensor networks: Attacks and countermeasures. *Ad hoc networks*, 1(2-3):293–315, 2003.
17. Ravi Kishore Kodali, Vishal Jain, Suvadeep Bose, and Lakshmi Boppana. Iot based smart security and home automation system. In *2016 international conference on computing, communication and automation (ICCCA)*, pages 1286–1289. IEEE, 2016.
18. Davit Marikyan, Savvas Papagiannidis, and Eleftherios Alamanos. A systematic review of the smart home literature: A user perspective. *Technological Forecasting and Social Change*, 138:139–154, 2019.
19. Gregory Mone. Intelligent living, 2014.

20. Kenichiro Noda. Google home: smart speaker as environmental control unit. *Disability and rehabilitation: assistive technology*, 13(7):674–675, 2018.
21. Olayemi Olawumi, Keijo Haataja, Mikko Asikainen, Niko Vidgren, and Pekka Toivanen. Three practical attacks against zigbee security: Attack scenario definitions, practical experiments, countermeasures, and lessons learned. In *2014 14th International Conference on Hybrid Intelligent Systems*, pages 199–206. IEEE, 2014.
22. Temitope Oluwafemi, Tadayoshi Kohno, Sidhant Gupta, and Shwetak Patel. Experimental security analyses of {Non-Networked} compact fluorescent lamps: A case study of home automation security. In *LASER 2013 (LASER 2013)*, pages 13–24, 2013.
23. Farhan Ar Rafi. *Application of Android and Raspberry Pi for Home Automation*. PhD thesis, JAHANGIRNAGAR UNIVERSITY, 2018.
24. Utkarsh Saxena, JS Sodhi, and Yaduveer Singh. Analysis of security attacks in a smart home networks. In *2017 7th International Conference on Cloud Computing, Data Science & Engineering-Confluence*, pages 431–436. IEEE, 2017.
25. Florian Schmeidl, Bara Nazzal, and Manar H Alalfi. Security analysis for smartthings iot applications. In *2019 IEEE/ACM 6th International Conference on Mobile Software Engineering and Systems (MOBILESoft)*, pages 25–29. IEEE, 2019.
26. R Sivapriyan, K Manisha Rao, and M Harijyothi. Literature review of iot based home automation system. In *2020 Fourth International Conference on Inventive Systems and Control (ICISC)*, pages 101–105. IEEE, 2020.
27. Shradha Somani, Parikshit Solunke, Shaunak Oke, Parth Medhi, and PP Laturkar. Iot based smart security and home automation. In *2018 Fourth International Conference on Computing Communication Control and Automation (ICCUBEA)*, pages 1–4. IEEE, 2018.
28. Satyendra K Vishwakarma, Prashant Upadhyaya, Babita Kumari, and Arun Kumar Mishra. Smart energy efficient home automation system using iot. In *2019 4th international conference on internet of things: Smart innovation and usages (IoT-SIU)*, pages 1–4. IEEE, 2019.
29. Pratik Waghmare, Prafull Chaure, Mayur Chandgude, and Abhay Chaudhari. Survey on: Home automation systems. In *2017 International Conference on Trends in Electronics and Informatics (ICEI)*, pages 7–10. IEEE, 2017.
30. T Watanabe, S Funakawa, T Kosaki, et al. Distribution and formation conditions of gibbsite in the upland soils of humid asia: Japan, thailand and indonesia. In *Proceeding of the 19th World Congress of Soil Science. Soil Solutions for a Changing World. Brisbane*, 2010.

Index

L. Fotia et al. (eds.), *Security, Trust and Privacy Models, and Architectures in IoT Environments*, Internet of Things, https://doi.org/10.1007/978-3-031-21940-5

The manufacturer's authorised representative in the EU is Springer Nature Customer Service Centre GmbH, Europaplatz 3, 69115 Heidelberg, Germany. If you have any concerns regarding our products, please contact ProductSafety@springernature.com

Printed and bound by CPI Group (UK) Ltd, Croydon, CR0 4YY
01/10/2026
02225875-0018